SECURING THE FUTURE OF U.S. AIR TRANSPORTATION

A System in Peril

Committee on Aeronautics Research and Technology for Vision 2050

Aeronautics and Space Engineering Board
Division on Engineering and Physical Sciences

Studies and Information Services
Transportation Research Board

NATIONAL RESEARCH COUNCIL
OF THE NATIONAL ACADEMIES

THE NATIONAL ACADEMIES PRESS
Washington, D.C.
www.nap.edu

THE NATIONAL ACADEMIES PRESS 500 Fifth Street, N.W. Washington, DC 20001

NOTICE: The project that is the subject of this report was approved by the Governing Board of the National Research Council, whose members are drawn from the councils of the National Academy of Sciences, the National Academy of Engineering, and the Institute of Medicine. The members of the committee responsible for the report were chosen for their special competences and with regard for appropriate balance.

This study was supported by Contract No. NASW-99037 between the National Academy of Sciences and the National Aeronautics and Space Administration. Any opinions, findings, conclusions, or recommendations expressed in this publication are those of the author(s) and do not necessarily reflect the views of the organizations or agencies that provided support for the project.

International Standard Book Number 0-309-09069-5 (Book)
International Standard Book Number 0-309-52738-4 (PDF)

Available in limited supply from the Aeronautics and Space Engineering Board, 500 Fifth Street, N.W., Washington, DC 20001, (202) 334-2858

Additional copies of this report are available from the National Academies Press, 500 Fifth Street, N.W., Lockbox 285, Washington, DC 20055; (800) 624-6242 or (202) 334-3313 (in the Washington metropolitan area); Internet, http://www.nap.edu

Printed in the United States of America

THE NATIONAL ACADEMIES

Advisers to the Nation on Science, Engineering, and Medicine

The **National Academy of Sciences** is a private, nonprofit, self-perpetuating society of distinguished scholars engaged in scientific and engineering research, dedicated to the furtherance of science and technology and to their use for the general welfare. Upon the authority of the charter granted to it by the Congress in 1863, the Academy has a mandate that requires it to advise the federal government on scientific and technical matters. Dr. Bruce M. Alberts is president of the National Academy of Sciences.

The **National Academy of Engineering** was established in 1964, under the charter of the National Academy of Sciences, as a parallel organization of outstanding engineers. It is autonomous in its administration and in the selection of its members, sharing with the National Academy of Sciences the responsibility for advising the federal government. The National Academy of Engineering also sponsors engineering programs aimed at meeting national needs, encourages education and research, and recognizes the superior achievements of engineers. Dr. Wm. A. Wulf is president of the National Academy of Engineering.

The **Institute of Medicine** was established in 1970 by the National Academy of Sciences to secure the services of eminent members of appropriate professions in the examination of policy matters pertaining to the health of the public. The Institute acts under the responsibility given to the National Academy of Sciences by its congressional charter to be an adviser to the federal government and, upon its own initiative, to identify issues of medical care, research, and education. Dr. Harvey V. Fineberg is president of the Institute of Medicine.

The **National Research Council** was organized by the National Academy of Sciences in 1916 to associate the broad community of science and technology with the Academy's purposes of furthering knowledge and advising the federal government. Functioning in accordance with general policies determined by the Academy, the Council has become the principal operating agency of both the National Academy of Sciences and the National Academy of Engineering in providing services to the government, the public, and the scientific and engineering communities. The Council is administered jointly by both Academies and the Institute of Medicine. Dr. Bruce M. Alberts and Dr. Wm. A. Wulf are chair and vice chair, respectively, of the National Research Council.

www.national-academies.org

COMMITTEE ON AERONAUTICS RESEARCH AND TECHNOLOGY FOR VISION 2050

RONALD R. FOGLEMAN, *Chair*, Durango Aerospace, Inc., Durango, Colorado
JACK CLEMONS, Lockheed Martin Air Traffic Management, Rockville, Maryland
WILLIAM B. COTTON, Flight Safety Technologies, Inc., Mount Prospect, Illinois
EUGENE E. COVERT, Massachusetts Institute of Technology, Cambridge
WILLARD J. DODDS, GE Aircraft Engines, Cincinnati
WILLIAM W. HOOVER, United States Air Force (retired), Williamsburg, Virginia
S. MICHAEL HUDSON, Rolls Royce North America (retired), Indianapolis
NANCY G. LEVESON, Massachusetts Institute of Technology, Cambridge
RICHARD MARCHI, Airports Council International–North America, Washington, D.C.
RICHARD R. PAUL, The Boeing Company Phantom Works, Seattle
AMY R. PRITCHETT, Georgia Institute of Technology, Atlanta
ROBERT J. RAVERA, RJR Aviation, LLC, Reston, Virginia
SANFORD REDERER, Aviation Planning and Finance, Arlington, Virginia
HERBERT H. RICHARDSON, Texas A&M University System, College Station
RUSSELL D. SHAVER III, RAND, Arlington, Virginia
DAVID D. WOODS, Ohio State University, Columbus

Staff

ALAN ANGLEMAN, Study Director
KARA BATH, Senior Project Assistant
BRIDGET EDMONDS, Senior Project Assistant
JENNIFER PINKERMAN, Research Associate
GEORGE LEVIN, Director, Aeronautics and Space Engineering Board

AERONAUTICS AND SPACE ENGINEERING BOARD

WILLIAM W. HOOVER, *Chair*, United States Air Force (retired), Williamsburg, Virginia
A. DWIGHT ABBOTT, Aerospace Corporation (retired), Los Angeles
RUZENA K. BAJSCY, NAE, IOM, University of California, Berkeley
JAMES BLACKWELL, Lockheed Martin Corporation, Marietta, Georgia
ANTHONY J. BRODERICK, Aviation Safety Consultant, Catlett, Virginia
SUSAN COUGHLIN, Aviation Safety Alliance, Washington, D.C.
ROBERT CRIPPEN, Thiokol Propulsion, Palm Beach Gardens, Florida
DONALD L. CROMER, United States Air Force (retired), Fallbrook, California
JOSEPH FULLER, Futron Corporation, Bethesda, Maryland
RICHARD GOLASZEWSKI, GRA Incorporated, Jenkintown, Pennsylvania
JAMES M. GUYETTE, Rolls-Royce, North America, Chantilly, Virginia
JOHN L. JUNKINS, Texas A&M University, College Station
JOHN M. KLINEBERG, Space Systems/Loral (retired), Redwood City, California
ILAN M. KROO, Stanford University, Stanford, California
JOHN K. LAUBER, Airbus North America, Inc., Washington, D.C.
GEORGE K. MUELLNER, The Boeing Company, Seal Beach, California
DAVA J. NEWMAN, Massachusetts Institute of Technology, Cambridge
JAMES G. O'CONNOR, NAE, Pratt & Whitney (retired), Coventry, Connecticut
MALCOLM R. O'NEILL, Lockheed Martin Corporation, Bethesda, Maryland
CYNTHIA SAMUELSON, Logistics Management Institute, McLean, Virginia
KATHRYN C. THORNTON, University of Virginia, Charlottesville
HANSEL E. TOOKES II, Raytheon International, Inc. (retired), Falls Church, Virginia
DIANNE S. WILEY, The Boeing Company, Long Beach, California
THOMAS L. WILLIAMS, Northrop Grumman, Bethpage, New York

Staff

GEORGE LEVIN, Director

Preface

In the past few years, the current status and future vision of the U.S. air transportation system have been examined in numerous studies. NASA's recent *Aeronautics Blueprint* notes that the United States and the world are becoming "more dependent on the ability to move goods and people faster and more efficiently by air. . . . Over the last century, aviation has evolved to become an integral part of our economy, a cornerstone of our national defense, and an essential component of our way of life. . . . Americans per capita use aviation more than any other country in the world, . . . [and nonbusiness] personal travel accounts for more than 50 percent of commercial air transportation." [1]

What is needed now is vigorous action to refine and achieve the broadly held future vision of an air transportation system that can meet consumer demands for safety, security, comfort, and convenience; public demands for environmental compatibility; and national economic demands for a globally competitive civil aeronautics industry. Achieving this vision will not be easy—and will not be possible without strong national leadership. Fortunately, sometimes the flow of history leads to a confluence of events that creates an opportunity to meet great challenges. As suggested by this committee in a letter report dated August 14, 2002,[2] the 100th anniversary of powered flight, which will take place in December 2003, provides an excellent opportunity both to create a bold new vision for air transportation and to initiate vigorous action by government agencies and private organizations to pursue that vision. Allowing this opportunity to pass without action would be a tragic mistake.

Ronald Fogleman, *Chair*
Committee on Aeronautics Research and Technology
for Vision 2050

---

[1]National Aeronautics and Space Administration (NASA). 2002. Aeronautics Blueprint. Available online at <www.aerospace.nasa.gov/aero_ blueprint/index.html>.

[2]National Research Council (NRC). 2002. Aeronautics Research and Technology for 2050: Assessing Visions and Goals—Letter Report. Washington, D.C.: National Academy Press. Available online at <www.nap.edu/catalog/10518.html>.

Acknowledgments

This report has been reviewed in draft form by individuals chosen for their diverse perspectives and technical expertise, in accordance with procedures approved by the Report Review Committee of the National Research Council (NRC). The purpose of this independent review is to provide candid and critical comments that will assist the institution in making its published report as sound as possible and to ensure that the report meets institutional standards for objectivity, evidence, and responsiveness to the study charge. The review comments and draft manuscript remain confidential to protect the integrity of the deliberative process. We wish to thank the following individuals for their review of this report:

Daniel Brand, Charles River Associates,
Jack E. Buffington, Mack-Blackwell National Rural Transportation Study Center,
Richard M. Carlson, Consultant,
Robert A. Davis, The Boeing Company (retired),
John J. Fearnsides, Lockheed Martin Air Traffic Management,
Gerald J. Iafrate, North Carolina State University,
Ilan Kroo, Stanford University,
Amedeo R. Odoni, Massachusetts Institute of Technology,
Clinton V. Oster, Jr., Indiana University,
Thomas B. Sheridan, Massachusetts Institute of Technology,
Agam N. Sinha, The MITRE Corporation,
Edmond L. Soliday, United Airlines (retired), and
Bill G.W. Yee, Belcan Corporation.

Although the reviewers listed above have provided many constructive comments and suggestions, they were not asked to endorse the conclusions or recommendations, nor did they see the final draft of the report before its release. The review of this report was overseen by Lester A. Hoel, University of Virginia, and Adib K. Kanafani, University of California, Berkeley. Appointed by the NRC, they were responsible for making certain that an independent examination of this report was carried out in accordance with institutional procedures and that all review comments were carefully considered. Responsibility for the final content of this report rests entirely with the authoring committee and the institution.

The committee also wishes to thank all those who supported the work of the committee, particularly Stephen Godwin, Director, Studies and Information Services, of the NRC's Transportation Research Board, and all those who participated in the committee's information-gathering meetings during Phase 1 or Phase 2, either in person or via teleconference:

Andy Anderegg, The MITRE Corporation
Gary Anderson, Army Research Office
Dale Ashby, Sikorsky
Doug Ball, The Boeing Company
Tom Berry, The MITRE Corporation
Alan Bloodgood, Lockheed Martin Air Traffic Management
James G. Boyd IV, Texas A&M University
Steve Brueck, University of New Mexico
Robert Buley, Northwest Airlines
Carl Burleson, Federal Aviation Administration
Philip Carrigan, Raytheon Air Traffic Management
Walt Coleman, Regional Airline Association (retired)
Sarah Dalton, Alaska Airlines
James W. Danaher, National Transportation Safety Board (retired)
Duane Dupon, Federal Aviation Administration
Igor Frolow, IBM Global Services
Pam Gernier, The MITRE Corporation
John R. Hansman, Jr., Massachusetts Institute of Technology
Pres Henne, Gulfstream Aerospace
Richard J. Hill, Air Force Research Laboratory
Urmila Hiremath, The MITRE Corporation
Gerald J. Iafrate, North Carolina State University
Siegfried Janson, Aerospace Corporation
Margaret Jenny, Consultant
John L. Junkins, Texas A&M University
Jack Kerrebrock, Massachusetts Institute of Technology
Alexander (Sasha) Klein, Preston Aviation Solutions
Larry Knolhoff, Department of Defense, Office of the National Security Space Architect
Peter Kostiuk, Logistics Management Institute
Ilan Kroo, Stanford University
Frederick Kuhl, The MITRE Corporation
Andrew Lacher, The MITRE Corporation
Richard Lareau, Transportation Security Administration
Timothy Lewis, Air Force Research Laboratory
Robert Liebeck, The Boeing Company
Sanford D. Mangold, Department of Defense, Office of the National Security Space Architect
Lourdes Maurice, Federal Aviation Administration
John McCarthy, Naval Research Laboratory
Jack McGuire, The Boeing Company
Robert E. McKinley, Jr., NASA Langley Research Center
Dennis A. Muilenburg, The Boeing Company
Raja Parasuraman, Catholic University of America
Paul Piscopo, Department of Defense
George Price, NASA Headquarters
Blaine Rawdon, The Boeing Company
Herm Rediess, Federal Aviation Administration
Othon Rediniotis, Texas A&M University
John Rekstad, Federal Aviation Administration
Harold Rosenstein, The Boeing Company
Karlin Roth, NASA Ames Research Center
Lillian Ryals, The MITRE Corporation
Marvin Schmidt, Universal Technology Corporation
Robert Schwab, The Boeing Corporation
Walt Smith, Pratt & Whitney

Ed Stevens, Raytheon Air Traffic Management
Jeffrey M. Stricker, Air Force Research Laboratory
Bob Vilhauer, The Boeing Corporation
Jim Walton, UPS Advanced Flight Systems
Fred Wieland, The MITRE Corporation
Rob Williams, Boeing Phantom Works
Richard Wlezien, NASA Headquarters
Ron York, Rolls-Royce North America
Rick Zelenka, The Boeing Corporation
Andres Zellweger, NASA Headquarters
Dorothy Zolandz, Nationa Research Council

Contents

APPENDIXES

Tables and Figures

TABLES

FIGURES

Executive Summary

As recently as the summer of 2001, many travelers were dreading air transportation because of extensive delays associated with undercapacity of the system. That all changed on 9/11, and demand for air transportation has not yet returned to peak levels.[1] Most U.S. airlines continue to struggle for survival, and some have filed for bankruptcy. The situation makes it difficult to argue that strong action is urgently needed to avert a crisis of undercapacity in the air transportation system. Yet that remains the case. History shows that crises of confidence, economic downturns, and international conflicts can depress the demand for air transportation, but only over the short term. In every earlier case, the long-term trend of increasing demand has reasserted itself. Assuming that current events have fundamentally and permanently changed the public's demand for air transportation is not a sound approach to preparing for the long-term future of the air transportation system. Current events have provided an opportunity for U.S. national leadership to create a comprehensive, widely accepted long-term vision and a coherent set of requirements from all federal agencies with a major stake in the air transportation system. The continued absence of a national-level endeavor to address the current situation threatens to place the air transportation system in increasing peril.

To help assure the future of the U.S. commercial air transportation system, the National Aeronautics and Space Administration (NASA) and the Federal Aviation Administration (FAA) requested that the National Research Council establish the Committee on Aeronautics Research and Technology for Vision 2050. The committee was charged with assessing (1) the visions and goals for U.S. civil aviation, as described in five key documents produced by the federal government,[2] and (2) technology goals for the year 2050. The committee issued a letter report on August 14, 2002, to address the first topic.[3] Current U.S. visions for civil aviation correctly point out the importance of civil aviation. To sustain our ability to reap the benefits that aviation provides, the U.S. visions consistently identify three main areas that long-term aeronautics research should address:[4]

- capacity of the air transportation system (in terms of passenger-miles, cargo-ton-miles, and aircraft operations)
- environmental compatibility (noise and emissions)
- safety and security

The committee concluded, however, that U.S. visions and goals consistently overlook several key items:

- a clear set of guiding principles
- a description of the overall process for developing and achieving a widely endorsed long-term vision for the air transportation system

[1]This report uses demand generally to refer to both consumer demand (the amount of air transportation services purchased, in terms of passenger-miles and cargo-ton-miles) and the load imposed on the air traffic control system (in terms of aircraft operations). Demand reflects the response of consumers to prices and the shape of the air transportation demand curve. Consumer demand is closely linked to demand on the air traffic control system, as individual airlines adjust routes, schedules, levels of service, prices, etc., to both stimulate and satisfy consumer demand.

[2]The complete statement of task appears in Appendix A, which also lists the visions assessed by the committee. A summary of the committee's comparative assessment appears in Appendix B.

[3]National Research Council (NRC). 2002. Aeronautics Research and Technology for 2050: Assessing Visions and Goals—Letter Report. Washington, D.C.: National Academy Press. Available online at <www.nap.edu/catalog/10518.html>.

[4]Items in this and other lists are either listed alphabetically or grouped topically. The committee did not prioritize research areas in each list.

- a strategy for overcoming transitional issues
- consumer satisfaction
- primacy of the U.S. aeronautics industry

Securing the future of the air transportation system requires that change within the system be accelerated quickly enough and directed with enough agility to avoid problems and achieve future goals while managing (1) the influence of increased demand and other external pressures and (2) conflicts between different goals and stakeholders. The process of achieving the long-term vision must be robust enough to prevent the system from changing too slowly, drifting, or going in the wrong direction.

The process of improving the long-term performance of the air transportation system—and organizing a corresponding long-term research and technology program—should start with a unified, widely endorsed national vision that specifies goals in each key area of interest to the commercial aviation community. The continued success of aviation and the benefits that it provides will require changes to accommodate increased demand. The committee found this to be the most critical long-term issue facing all aspects of the air transportation system. Issues associated with safety, security, and environmental compatibility are also exacerbated by greater demand, and the effectiveness of currently envisioned near-term solutions in each of these areas would be diminished if demand for air travel in the United States doubles over the next 10 to 35 years, as currently projected. Increasing passenger throughput enough to keep up with increased demand requires eliminating constraints and improving the flexibility of the system enough to overcome localized capacity problems while accommodating the full range of authorized users (commercial, private, and military). For example, eliminating the effects of adverse weather is not enough; in many areas, the baseline capacity of the system (in good weather) must also be greatly increased to accommodate a deregulated airline industry as it strives to meet user demands for convenient service. This requires research leading to improvements in every element of the air transportation system.

The future vision of the air transportation system should be supported by research and technology goals leading to improved performance in terms of en route comfort of passengers, the convenience of passenger travel and air freight service (including travel time), the cost of moving passengers and cargo (including the cost of developing and manufacturing new aircraft and aircraft systems), and the societal impact of aviation (in terms of the consumption of nonrenewable fuels, emissions, land use, noise, safety, security, reduced congestion in other modes of transportation, employment, and other effects on the national economy). Measurable long-term targets supported by sound analyses should be established to assess progress toward the goals. Research should support the establishment of quantifiable goals in areas where progress is difficult to measure.

The air transportation system is supported by a core of dedicated government and industry personnel who are developing new operational concepts, architectures, and modernization plans. Yet no single organization has the responsibility and authority for developing a comprehensive solution to the challenges faced by the U.S. air transportation system. Business as usual, in the form of continued, evolutionary improvements to existing technologies, aircraft, air traffic control systems, and operational concepts, is unlikely to meet the needs of air transportation over the next 25 to 50 years. The disparity between (1) the rate at which demand is increasing and (2) the rate at which technology is reducing aircraft noise and emissions is becoming increasingly difficult to overcome because technical advances are becoming increasingly difficult to achieve. Without strong, focused leadership, the likely result will be an air transportation system where growth in demand has been greatly curtailed by undercapacity; the environmental effects of aviation; customer dissatisfaction with available levels of comfort, convenience, and cost; and/or factors related to safety and security.

The committee believes that strong action by a federal agency or office to provide such leadership, with the broad support of the administration and the Congress, would do more to improve the ability of national aeronautical research and development programs to achieve their goals than any other change in the management or content of the programs themselves. The designated office should have (1) the responsibility, authority, and financial resources necessary for defining air transportation system architectures through a centralized planning function, (2) an understanding of the interactions among system performance parameters, demand, and economic factors, such as the methods used to fund federal activities in support of the air transportation system, and (3) the credibility and objectivity to garner the active support of other air transportation stakeholders in government, industry, and the general public. This will require, among other things, a leadership group composed of individuals with a broad aviation perspective and a willingness to accept the risks of looking ahead and allowing others to help define the future.[5]

PROCESS FOR CHANGE

The aviation system is unique in that it has one federal agency (NASA) responsible for long-range research and development and another agency (FAA) that supplies traffic management systems and services and regulates the carriers and manufacturers. The cultures, missions, and operating practices of NASA's aeronautics enterprise and the FAA are

[5]Assessing the organization and role of specific government agencies was beyond the scope of this study (see Appendix A), so no recommendation is made regarding which federal office or agency should be designated to provide the required leadership.

quite distinct, as would be expected when comparing a research organization with an operational organization. Nonetheless, they are the federal government's principal agents for operating and improving the technical capabilities of the air transportation system.

A national vision, clear technology goals, and strong, focused leadership are necessary to improve the competitiveness of the U.S. aeronautics industry and enable the air transportation system to satisfy increased demands for air travel without degrading system safety, security, environmental compatibility, or consumer satisfaction. Also required is a process for integrating, organizing, and directing aeronautics research and technology development and a clear understanding of organizational roles. Action necessary to achieve the above is encapsulated in the process for change that is defined in the following summary recommendation:

Recommendation. Process for Change. Establish air transportation as a national priority with strong, focused leadership. Air transportation system technology planning and development should be done in the context of a process driven by the needs of system users and the nation as a whole.

1. Implement a public/private process for change, as follows:
   - Designate a federal agency or office to provide strong leadership in overcoming the challenges faced by the U.S. air transportation system.
   - Establish an interagency process for developing and achieving a widely endorsed long-term vision of the air transportation system that includes a clear set of guiding principles and a strategy for overcoming transitional issues.
   - Document the process.
   - Coordinate action and resolve disputes among stakeholders in the aviation community with different concerns and priorities (e.g., manufacturers and operators; executives and employees; pilots, controllers, and passengers; local, federal, and state governments; regulators; the military; and general aviation).
   - Gather and analyze feedback on how well the process is working from the perspective of all interested parties, especially when conditions change, to identify problems before serious incidents or disruptions occur and to recognize new opportunities.
   - Formally review the process and process outputs at least every 4 years.
   - Update the process.
2. The output of the process should include the following:
   - A better understanding of future demand for air transportation to make sure that changing trends will be detected as soon as possible.
   - A unified, long-term national vision endorsed and supported by the aeronautics community as a whole and cognizant federal agencies.
   - Broad public policies to support the vision.
   - Long-term operational concepts to meet the vision and to serve as a continuing resource for guiding change and coordinating action by different parties.
   - System architectures to realize the operational concepts.
   - An understanding of how the U.S. air transportation system of the future will fit into the national (intermodal) transportation system and the international air transportation system.
   - Validated research and technology requirements.
   - An implementation plan to achieve all of the above, including a clear understanding of government and industry roles in developing precompetitive and noncompetitive aeronautical research and transitioning the results of civil and military government research to commercial development.
3. A comprehensive suite of system models should be developed, validated, and maintained to support informed decision making throughout the process. Models should encompass the following:
   - demand
   - economics
   - environmental effects
   - existing and new technologies
   - human performance
   - interactions with other modes of transportation
   - new operational concepts
   - organizational factors
   - security threats and preventive measures
   - system engineering
   - transition (from old to new technologies, systems, and organizational structures)
4. A commitment should be made to support a stable long-term research program to provide the knowledge, tools, and technologies needed throughout the process. At a low level, the research program should investigate innovative research ideas that challenge accepted precepts.

The following sections describe in more detail specific actions for improving the performance of (1) the air transportation system as a whole, (2) modeling and simulation capabilities necessary to support improvements in the air transportation system, and (3) individual aircraft.

IMPROVING THE AIR TRANSPORTATION SYSTEM

Developing meaningful and useful operational concepts stemming from a broadly defined vision of the air transportation system 25 to 50 years hence is a critically important task in the process of improving the performance of the system. To meet this challenge, the federal government, working with other stakeholders in the air transportation system, should develop a coherent set of operational concepts supporting a vision of the air transportation system in the 2050 time frame. These concepts should encompass a range of potential changes in technology, society, and the air transportation system itself. They should be used to guide long-term research and the evolution of and transition to a more advanced air traffic management system. The concepts should be continually, objectively, and rigorously evaluated (for example, through comprehensive simulation and modeling) and iterated to reflect feedback from stakeholders, conflicts between alternative concepts, and the best understanding of the future costs, benefits, and requirements that are likely to evolve in response to changes in the real world, the current state of technology and systems operations, and future expectations.

The research and technology requirements should be tailored to meet the requirements of future operational concepts. Enabling technologies applicable to a wide range of operational concepts should be developed in parallel with research to develop and evaluate long-term operational concepts so that the necessary technologies will be ready for whichever operational concept proves to be most beneficial. Technology areas of particular interest include the following:

- design of human-integrated systems
- distributed, collaborative decision making
- autonomous and interactive technologies
- noise and emissions locally, regionally, and globally
- wake vortices
- situational awareness
- systems-engineering methods
- avionics

Technological research alone is insufficient to achieve the future vision. Research is also needed to (1) better understand the economic, environmental, political, institutional, and managerial factors involved in achieving key goals, (2) take advantage of synergies among these factors, and (3) overcome related impediments. The federal government should support research to develop improved processes and methods in the following nontechnology areas:

- economics
- regulations, certification requirements, and operating procedures
- resolution of conflicting objectives of different stakeholders
- societal concerns about aircraft noise and emissions

MODELING AND SIMULATION

Federal, industrial, and academic institutions in the United States have tremendous research capabilities and resources. Achieving the future vision of the air transportation system requires that research be directed at technical capabilities most likely to achieve long-term performance goals. Complementary use of field tests, laboratory tests, modeling, analysis, and simulation would improve the ability to (1) measure systemwide behavior of the air transportation system, (2) assess the performance of proposed operational concepts, technologies, and other changes, and (3) make informed investment decisions to reduce the schedule, cost, and technical risks of system improvements. In addition, the process of securing the future would be greatly facilitated if the federal agencies that support research in aviation system models would improve their coordination, especially with regard to the following:

- research plans
- participation of industry and academia
- criteria for maintenance and validation
- availability of models
- use of models by decision makers

The government and other interested parties should support additional simulation and modeling research in the following areas:

- interoperability
- safety analysis
- demand and demand allocation
- validation of models and suites of models
- formation of a suite of system models
- role of humans in the aviation system of the future

IMPROVING AIRCRAFT PERFORMANCE

Improvements in aircraft performance are critical to achieving necessary improvements in almost every aspect of the overall performance of the air transportation system. Innovative long-range research leading to the implementation of new operational concepts is also required for the air transportation system to take full advantage of gains in the performance of commercial aircraft.

To improve the performance of aircraft through 2025, federal agencies should continue to support research leading to evolutionary improvements in aircraft performance. Looking out to 2050, however, large gains in aircraft performance are unlikely to be achieved without innovative long-range research leading to new aircraft concepts and technologies. Areas of particular interest include the following:

- analytical tools
- composite materials
- environmental consequences of aircraft noise and emissions
- low emissions combustor technology
- nanotechnology
- nontraditional aircraft configurations

- nontraditional power and propulsion concepts
- passive and active control of laminar and turbulent flow
- high-temperature engine materials and advanced turbomachinery

Technologies specifically related to personal air vehicles, uninhabited air vehicles, supersonic aircraft, or runway-independent air vehicles have the potential to improve the performance of the air transportation system, especially in niche areas. However, research in these areas will not be able to resolve the overall capacity problems that are the primary challenge to the continued success of the air transportation system over the long term. Accordingly, the committee did not examine technologies related to these vehicle classes and makes no recommendations concerning the future direction of research in these areas. Nonetheless, the process for change recommended by the committee would facilitate the planning of research for all vehicle types.

1

Foundation for Change

At the request of the National Aeronautics and Space Administration (NASA) and the Federal Aviation Administration (FAA), the National Research Council (NRC) established the Committee on Aeronautics Research and Technology for Vision 2050 to assess (1) the long-term visions and goals for U.S. civil aviation, as described in five key documents produced by the federal government, and (2) technology goals for the year 2050 (see Appendix A). The committee issued a letter report on August 14, 2002, to address the first topic.[1] The substance of that letter has been incorporated into this report, which also addresses the second topic.

This chapter describes key elements of the future vision and the need for specific goals to support that vision. After acknowledging the limits of any effort to look far into the future, the chapter then describes the primary challenge to achieving the future vision and previews a process for making needed changes. Three subsequent chapters address research related to improving (1) air transportation system performance, (2) system-level modeling of the air transportation system, and (3) performance of individual aircraft. The final chapter concludes with a summary recommendation regarding the process for change that is vital to securing the future of the air transportation system. A complete list of the findings and recommendations contained in this report then follows, along with appendixes that contain the statement of task and the study approach executed by the committee, a comparative assessment of future goals and visions, brief biographies of the members of the committee, a descriptive catalog of propulsion system concepts, and a description of four levels of system models.

[1]National Research Council (NRC). 2002. Aeronautics Research and Technology for 2050: Assessing Visions and Goals—Letter Report. Washington, D.C.: National Academy Press. Available online at <www.nap.edu/catalog/10518.html>.

VISION AND GOALS

To continue to reap the benefits that the air transportation system provides, the U.S. visions examined by the committee consistently identify three main thrusts that long-term aeronautics research should address: safety and security, capacity, and environmental compatibility (noise and emissions). At the same time, the U.S. visions and goals consistently overlook several key items: a description of the overall process for developing and achieving a widely endorsed long-term vision for the air transportation system, a clear set of guiding principles, and a strategy for overcoming transitional issues.

In assessing the U.S. goals and visions, the committee also examined a comparable vision for civil aeronautics in Europe. The European vision highlighted two additional areas that are missing from the U.S. visions. The latter do not include as a goal the satisfaction of consumer needs—that is, the quality and affordability of air transportation—perhaps because consumers do not seem to have been consulted when the U.S. visions were formulated. This could be a major oversight, given the large role that consumer demand for low cost and convenience (e.g., frequent departures) plays in business decisions made by industry.[2] Also, although the U.S. visions as a whole recognize that national well-being depends on a national transportation system with a strong

[2]This report uses demand generally to refer to both consumer demand (the amount of air transportation services purchased, in terms of passenger-miles and cargo-ton-miles) and the load imposed on the air traffic control system (in terms of aircraft operations). Demand reflects the response of consumers to prices and the shape of the air transportation demand curve. Consumer demand is closely linked to demand on the air traffic control system, as individual airlines adjust routes, schedules, levels of service, prices, etc., to both stimulate and satisfy consumer demand.

aviation element, they do not include primacy of the U.S. aeronautics industry as a goal. Competitiveness is so central to the European vision, by contrast, that it appears in the title of the document that defines this vision: *European Aeronautics: A Vision for 2020—Meeting Society's Needs and Winning Global Leadership.*

The vision for the U.S. air transportation system should be supported by research and technology goals leading to improved system performance. Measurable long-term targets should be established to assess progress toward those goals. Ideally, goals should be

- Ambitious enough to be challenging, without going beyond the limits of what is practical given likely constraints imposed by the current and future state of scientific and engineering knowledge, economics, and other nontechnical factors.
- Linked to specific benefits—for example, noise goals that will end the exposure of communities near airports to a day-night average sound level greater than 55 dB.
- Focused on areas in which government research can have a direct impact.
- Supported by research that demonstrates that the goals will result in the intended outcome—for example, that noise will be substantially eliminated as a constraint on airport operations.
- Time-phased, with different levels of performance established over different periods of time where appropriate.

Organizational goals should be dynamic to respond to a changing world and changing requirements. NASA recently replaced time-phased, quantitative goals for aeronautics research and technology (e.g., "double the aviation system capacity within 10 years, and triple it within 25 years") with open-ended, qualitative goals (e.g., "enable more people and goods to travel faster and farther, anywhere, anytime, with fewer delays"). Quantifiable goals may be difficult to establish for research leading to improved understanding and for research related to customer satisfaction, competitiveness, and improving human-computer interactions. However, limiting research to areas with easily quantifiable goals would reduce the scope of research to a subset of the overall problem. Furthermore, quantifiable goals could be readily established in many areas, and even as NASA moves away from quantitative research goals, the report of the Commission on the Future of the U.S. Aerospace Industry (2002) recommends the adoption of specific, quantifiable aerospace technology demonstration goals for capacity, safety, mobility, and environmental effects as a national priority.

In considering how research could improve the performance of the air transportation system, the committee took a broad view of performance that considers the particular needs of customers, airlines, and manufacturers. This broad view includes the following parameters (listed alphabetically):

- Comfort en route
  - —en route accommodations
  - —transfer activities, including airport amenities
- Convenience of passenger travel and air freight service
  - —availability of service at times desired by customers
  - —availability of service to desired departure and destination locations
  - —ease of passenger ingress and egress, cargo handling, and aircraft handling, especially as it relates to customer satisfaction and capacity constraints
  - —total travel time
- Cost of moving passengers and cargo
  - —cost of developing and manufacturing new aircraft and aircraft systems
  - —cost of passenger and freight operations
  - —ticket prices for passenger travel and prices paid for freight services
- Societal impact
  - —consumption of nonrenewable fuels
  - —effects on the national economy (employment, etc.)
  - —emissions
  - —land use
  - —noise
  - —reduced congestion in other modes of transportation
  - —safety and security

Airline economics are complex, and the relationship between technology and some of the above parameters is indirect and difficult to quantify. For example, technological efficiencies will not be able to compensate for the economic inefficiencies that occur when reduced demand (or a poorly structured route system) produces low load factors, resulting in high costs per passenger-mile (or cargo-ton-mile). Furthermore, in the short term, ticket prices are often driven more by competitive pressures and the laws of supply and demand than by the productivity and efficiency of the aircraft used to provide transportation services. Ultimately, however, improved performance (that results in lower costs) is necessary because prices that fall below the service providers' average total costs are economically unsustainable without external subsidies.

Recommendation 1-1. Goals. The future vision for the air transportation system should be supported by research and technology goals leading to improved performance. Measurable long-term targets supported by sound analyses should be established to assess progress toward the goals. Research should support the establishment of quantifiable goals in areas where progress is difficult to measure.

BEYOND THE HORIZON

When considering the future of the U.S. air transportation system, the committee sees a fundamental difference in expectations between (1) the next 25 years or so and (2) the quarter century ending in 2050. Commercial aircraft and air traffic management systems have lifetimes on the order of 25 years. In addition, the commercial aviation industry—as well as the government's certification processes—are justifiably cautious when it comes to the acceptance of new concepts and technologies. As a result, the gestation period between initial development of a new technology or concept and the point where it sees widespread operational use is often measured in decades. Accordingly, the state of the air transportation system in 2025 will largely be defined by systems that are already operational and the implementation of existing research. In other words, 2025 is imaginable in terms of current technology. The air transportation system of 2050, however, could incorporate technologies that are as yet undiscovered or that exist but will be used in ways as yet unimagined. However, the air transportation system of 2050 will benefit from new ideas only if long-term research—and the technology recommendations made in subsequent chapters of this report—are pursued with enough vigor to ensure that the long gestation associated with new technological approaches and operational concepts will be completed by then.

Looking backward to 1953 illustrates the challenge of predicting the future 50 years hence. The personal computer had not yet been invented in 1953, yet any 50-year prediction of aviation requirements issued in 1953 that did not account for this unforeseen technology would look absurd today. What new technologies will take aviation in unexpected directions over the next 25 to 50 years? An authoritative answer is impossible, so long-range plans must be flexible enough to accommodate the unexpected. For decades, railroads were the preeminent long-distance transportation system. But the U.S. rail system of today is vastly different from the system of the early 1900s. Railroads are still a dominant provider of cargo service, yet railroads are now a minor provider of long-distance passenger service, and the passenger service that survives has done so only because the government subsidizes operational and capital costs. For all transportation modes flexibility in the face of changing priorities and challenges is essential.

CHALLENGE

Current U.S. visions for civil aviation correctly point out the importance of civil aviation to the national economy and overall standard of living. People want to travel quickly and comfortably. Businesses and their customers want products delivered overnight. Per capita use of aviation is higher in the United States than any other country in the world, and nonbusiness travel accounts for more than 50 percent of passenger air travel. The availability of quick and affordable options for long-distance travel increases demand. That will probably always be the case. To the extent that air transportation can continue to satisfy these universal human desires safely, reliably, and affordably, air transportation will remain relevant.

Yet many people do not enjoy air travel. Technologies associated with videoconferences, netconferences, and virtual reality experiences are gradually increasing the ability to be "present" without being there. Crises such as the shutdown of the air transportation system following 9/11, security concerns that still keep many air travelers at home, security procedures that increase travel time, international conflicts, a downturn in the national and global economies, and the SARS epidemic are providing incentives for the corporate world to find new ways of doing business remotely. The same concerns are also inducing many leisure travelers to plan vacations that do not require travel by air.

Though the committee does not believe that current problems are a harbinger of a change in long-term trends, changes in the way customers use the air transportation system—and changes in the way government and industry structure and use the air transportation system—need to be anticipated wherever possible. For example, a significant shift of airspace use away from business travel to a broad mix of nonbusiness travel and cargo—even as overall demand continued to increase—would have significant impact on service providers and the air traffic control system. Missing such a trend could result in a misdirection of research.

Two summers ago the air transportation system experienced extraordinary delays as the demand for air travel pushed traffic to the limit and disrupted schedules in many parts of the system. Although the air transportation system had faced extensive delays before, the crisis two summers ago was quantitatively the gravest in history. Since then, the demand for air transportation has been curtailed by the factors noted above. Today, undercapacity is no longer an immediate problem in most parts of the U.S. air transportation system, and air transportation faces the peril of economic devastation as major carriers reduce service and file for bankruptcy. Although the committee believes that vigorous action is needed to prevent capacity problems in the years ahead from producing unacceptable delays in air transportation, the future is by nature uncertain. But some things are certain. Agility in responding to changing situations and trends is vital. National leadership is essential for motivating and directing change. A robust suite of systems models is needed to explore future possibilities and secure the future regardless of what eventually comes to pass. Lastly, uncertainty about the future must not stand in the way of aggressive action to address the problems that *are* expected. The only alternative is to abandon the opportunity to get ahead of future problems and simply react to crises as they occur.

In the past, short-term downturns in demand for air transportation have always been followed by a return to steady growth in demand, and the committee believes that this pattern will repeat itself. As a result—despite the current situation—the most critical issue facing all aspects of the air transportation system over the long term is likely to be growth in demand for air transportation. To increase capacity, the system can be expanded and the capabilities of the system can be increased, but many elements of the air transportation system—including many major airports—are constrained. In some areas, significantly expanding the system infrastructure would be expensive, time-consuming, and extraordinarily difficult. Issues associated with safety, security, and environmental compatibility are also exacerbated by greater demand, and the effectiveness of currently envisioned near-term solutions in each of these areas would be diminished if demand for air travel in the United States doubles over the next 10 to 35 years, as currently projected.[3] Air transportation leadership must remain mindful of changes, both within and without the air transportation system, and remain flexible if the system is to remain vital over the long run.

Finding 1-1. Challenge of Increased Demand. The continued success of aviation and the benefits that it provides will require changes to accommodate increased demand. This is the most critical long-term issue facing all aspects of the air transportation system. Issues associated with safety and security, capacity, environmental compatibility, and consumer satisfaction are all exacerbated by greater demand, and the effectiveness of near-term solutions in each of these areas will be diminished by long-term growth in demand for air transportation in the United States.

Business as usual, in the form of continued, evolutionary improvements to existing technologies, aircraft, air traffic control systems, and operational concepts, is unlikely to meet the challenge of greatly increased demand over the next 25 to 50 years. The importance of altering historical trends is particularly important to limit the environmental effects of aviation. Between 1976 and 2001, the demand for air transportation increased by 250 percent, yet the fuel efficiency (amount of fuel consumed per passenger-mile or ton-mile of cargo) of large commercial aircraft in the United States, which reflects improvements in the design of both the airframe and the propulsion system, improved by only 50 percent. As a result, air transportation is burning more fuel and producing more emissions, which contribute to environmental problems locally and globally. Fuel consumption and engine emissions have not gone up as much as they would have without new technology, but they are going up nonetheless. A similar situation exists with noise. Although each new generation of aircraft produces less noise than older aircraft of the same size, aircraft noise is still an unwelcome part of life in many airport communities, and limits on aircraft noise continue to constrain aircraft operations at many airports.

The disparity between (1) the rate at which demand is increasing and (2) the rate at which technology is reducing aircraft noise and emissions is becoming increasingly difficult to overcome because technical advances are becoming increasingly difficult to achieve. For example, the rate of improvement in specific fuel consumption has long been predicted to diminish as turbojet engines approach their theoretical limits (Dawson, 1968). NASA's now obsolete goals for aeronautics research recognized the importance of accelerating the rate of technological advances. Meeting the noise goals adopted by NASA in 1997 would have required a dramatic break from historical trends (NRC, 2002), but NASA has replaced those goals with new research goals that lack any measurable targets.

Revolutionary changes are needed in more than just aircraft and aircraft systems. Much of the current effort to increase system capacity is focused on eliminating delays caused by specific constraints, such as restricted visibility or other forms of adverse weather en route, in the terminal area,[4] and at airports. The problem faced over the 2050 time frame, however, is quite different. To increase passenger throughput enough to keep up with increased demand, eliminating the effects of adverse weather is not enough; in addition, the baseline capacity of the system in good weather must also be greatly increased. This may require widespread adoption of operating concepts that use runways and airspace in new ways. It may also require new paradigms for the air traffic management system as a whole, to leverage the significant advances that are being made in information technology and global surveillance, communication, and navigation capabilities.

Travel by air in the United States is extraordinarily safe, and industry and government make a tremendous effort to keep it so. Increased demand over the next 25 to 50 years could result in more accidents. More traffic could stress overloaded portions of the system to the point where accident rates increase, and even if the accident rate stayed the same, more traffic would result in more accidents per year. Historically, however, safety improvements have been able to reduce the total annual number of fatalities from commercial aircraft accidents despite increased demand. In fact, during 2002 U.S. airlines experienced no fatal accidents.

Future changes to the air transportation system to increase

[3]Forecasts of future demand by the government, industry, and other organizations and individuals predict that air travel will double in the next 10 to 35 years (Boeing, 2000; Neufville, 2000; Federal Transportation Advisory Group, 2001; NASA, 2002; NRC, 2002; RTCA Free Flight Steering Committee, 2002; FAA, 2003). Although the rate of increase is uncertain, all agree that air traffic operations will increase substantially over the long term.

[4]Terminal areas include the airspace within about 50 miles of major airports.

capacity could conceivably create unexpected hazards that lead to a higher accident rate. However, accidents are so unacceptable that such hazards would be corrected as quickly as possible, even if it meant undoing capacity enhancements. A similar philosophy applies to security, in the sense that security measures take precedence over capacity concerns. The primary challenge for security-related technologies is to increase security without constraining capacity. As with the other major thrusts, long-term plans for developing technology to improve security should be based on a systematic approach that assesses the specific problems that need to be addressed and targets technologies accordingly. For example, security systems should be designed to predict and adapt to future threats to ensure that we are not in a constant state of preparing to "fight the last war." Achieving this goal is difficult, because the foreseeable future can change in the instant that tragedy strikes in the form of a previously unforeseen failure of existing safety and security systems.

Addressing future challenges is also complicated because isolated efforts to achieve one goal may make it more difficult to achieve other goals. In addition, different stakeholders in the aviation community (manufacturers and operators; executives and employees; pilots, controllers, and passengers; local, federal, and state governments; regulators; the military; general aviation; and others) have different priorities. For example, everyone is in favor of reduced environmental impacts. However, passengers are very price conscious, and airlines that consistently fail to make a profit ultimately cease to exist. Therefore, advanced technology that reduces engine emissions but increases costs will be difficult to sell to airlines that are already meeting regulatory standards—unless those standards are expected to become more stringent.

Finding 1-2. Going Beyond Business as Usual. Business as usual, in the form of continued, evolutionary improvements to existing technologies, aircraft, air traffic control systems, and operational concepts, is unlikely to meet the needs for air transportation that will emerge over the next 25 to 50 years. The likely result would be an air transportation system where growth in demand has been greatly curtailed by undercapacity in the air traffic management system; the environmental effects of aviation; customer dissatisfaction with available levels of comfort, convenience, and cost; and/or factors related to safety and security.

The Big Question. How can change within the air transportation system be accelerated quickly enough and directed with enough agility to avoid problems and achieve future goals while managing (1) the influence of increased demand and other external pressures and (2) conflicts between different goals and stakeholders? How can the system be prevented from changing too slowly, drifting, or going in the wrong direction?

The answer is to develop an improved process to guide and facilitate change. Such a process is discussed in more detail below and in Chapter 5, which presents the committee's recommendation to institute an effective process for change.

CHANGE

The process of organizing a long-term research and technology program for civil aviation should start with a systematic statement of the underlying problems and a unified national vision to ensure that efforts by individual departments and agencies of the federal government respond to these problems in a synergistic fashion. Currently, however, most of the five vision documents examined by the committee have not been endorsed by the heads of the agencies who chartered them, and they contain goals that are inconsistent with the research and acquisition budgets of the federal agencies responsible for aviation. The situation raises questions about the relevancy of existing visions and demonstrates the need for federal agencies involved in civil aeronautics research and technology to support and implement a unified national vision.

The Department of Transportation has primary responsibility for civil aviation policy and regulation. Through the FAA, the Department of Transportation also has purview over the certification of civil aviation equipment and personnel, development and operation of the air traffic management system, and system safety. The functions and vital interests of many other government agencies are also related to the development and operation of the U.S. civil air transportation system. The federal government recently established a joint aviation systems program office involving the FAA, NASA, the Department of Commerce, the Department of Homeland Security, the Department of Defense, and other federal agencies. This office has the potential to enhance interagency cooperation in supporting the modernization of the air transportation system.

Operational concepts can help to develop the functional requirements for a new system by describing how it will perform, including the allocation of roles and responsibilities to interconnected systems and humans. Operational concepts may include system development, production, deployment, training, operation, maintenance, upgrading, and decommissioning. For example, the operational concept for an air traffic management system that allows parallel operations on closely spaced runways in low visibility conditions would describe approaches for dealing with safety concerns (e.g., wake vortices and collision), roles and responsibilities of pilots and controllers, equipment requirements (for airports and aircraft), and regulatory changes. At a higher level, operational concepts can be used to suggest how new kinds of aircraft, air traffic management procedures, systems, regulations, and business practices could improve the performance of the air transportation system.

New technologies and operational concepts should be assessed in terms of their ability to solve the key problems of the air transportation system of the future. Such an assessment requires modeling at the system level how the air transportation system is affected by various technologies, operational concepts, and external factors (e.g., economic conditions, intermodal travel options, and the future state of the international air transportation system). To facilitate the planning of long-term research, models should be able to assess the impact of proposed technologies and operational concepts beginning at the earliest stages of development. The assessments should also be focused on the overall performance of the air transportation system rather than on individual parameters or components of the system. For example, in the next 50 years it will probably become technologically feasible to replace pilots and/or air traffic controllers with automated systems, at least under nominal operating conditions. But to what extent would such an approach solve the key problems of today, and what new problems might such a solution introduce, especially during emergencies? The guiding principle should be to design synergistic partnerships between humans and automation that result in better performance in all operating conditions than either could achieve alone, rather than the false goal of trying to replace humans with computers.

Instead of beginning with the current state of the air transportation system, which could be appropriate for defining short-term goals, the development of long-term goals and visions should start by defining systemwide functional and performance requirements. The desired future state of the air transportation system, as one element of a multimodal national transportation system, should be defined using a comprehensive architecture that combines process elements for each dimension of the system (operational, system, technical, and economic). The future vision should also consider transitional issues:

- An environment that is conducive—in terms of regulations, regulatory approval processes, the certification process, operational procedures, and the perceptions of system operators, the traveling public, and society at large—to the introduction of new technologies and operational concepts.
- Interim improvements to the air transportation system along the way to the future.
- Incentives for government agencies and private industry to cooperate in defining and achieving a common vision.

The vision should be dynamic, able to change over time as societal needs, global events, and advances in technology alter the perception of what is desirable and possible.

Developing a comprehensive, unified vision for the future of the U.S. air transportation system—and generating widespread support to achieve the vision—will be a tremendous challenge. No single organization has responsibility for developing solutions that encompass the economic performance of private and governmental service providers, safety, security, environmental effects, and consumer satisfaction. Little is likely to happen without air transportation being clearly established as a national priority with strong, focused leadership. In fact, the committee believes that strong action by a federal agency or office to provide such leadership, with the broad support of the administration and the Congress, would do more to improve the ability of national aeronautical research and development programs to achieve their goals than any other change in the management or content of the programs themselves.[5]

Finding 1-3. Context for Future Requirements. Valid research requirements for the air transportation system depend on understanding how the U.S. air transportation system of the future will fit into both the national (intermodal) and international air transportation systems.

Recommendation 1-2. National Vision. The process of improving the long-term performance of the air transportation system—and organizing a corresponding long-term research and technology program—should start with a unified, widely endorsed, national vision that specifies goals in each key area of interest to the commercial aviation community. The vision should establish goals related to safety and security, the capacity of the air transportation system, environmental compatibility (noise and emissions), the satisfaction of consumer needs, and industrial competitiveness. It should include a clear set of guiding principles and a strategy for overcoming transitional issues.

Recommendation 1-3. Leadership. No single organization has the responsibility and authority for developing a comprehensive solution to the challenges faced by the U.S. air transportation system. Strong, focused leadership is needed. Federal leadership should be exercised by an agency or office with (1) the responsibility, authority, and financial resources necessary for defining air transportation system architectures through a centralized planning function, (2) an understanding of the interactions among system performance parameters, demand, and economic factors, such as the methods used to fund federal activities in support of the air transportation system, and (3) the credibility and objectivity to garner the active support of other air transportation stakeholders in government, industry, and the general public. This requires, among other things, a leadership group composed of individuals with a broad aviation perspective and a willingness to accept the risks of (1) looking ahead and (2) allowing others to help define the future.

---

[5]Assessing the organization and role of specific government agencies was beyond the scope of this study (see Appendix A), so no recommendation is made as to which agency or office should be designated to provide the required leadership.

REFERENCES

Boeing. 2000. World Market Outlook. Seattle, Wash.: The Boeing Company.

Commission on the Future of the U.S. Aerospace Industry. 2002. Final Report of the Commission on the Future of the U.S. Aerospace Industry. Washington, D.C.: U.S. Department of Commerce, International Trade Administration, Office of Aerospace. Available online at <www.aerospacecommission.gov/AeroCommissionFinalReport.pdf>.

Dawson, L. 1968. Propulsion. Aeronautical Journal of the Royal Aeronautical Society 72:209–229.

Federal Aviation Administration (FAA). 2003. FAA Aerospace Forecasts. Selected Aviation Demand Measures, Calendar Years 2003-2014, Table I-9. Washington, D.C.: FAA Office of Aviation Policy and Plans. Available online at <http://apo.faa.gov/foreca02/2003tab/exsumtableI-9.xls>.

Federal Transportation Advisory Group. 2001. Vision 2050: An Integrated Transportation System. Washington, D.C.: Department of Commerce. Available online at <http://scitech.dot.gov/polplan/vision2050/index.html>.

National Aeronautics and Space Administration (NASA). 2002. Aeronautics Blueprint. Washington, D.C.: NASA. Available online at <www.aerospace.nasa.gov/aero_blueprint/index.html>.

National Research Council (NRC). 2002. For Greener Skies: Reducing Environmental Impacts of Aviation. Washington, D.C.: National Academy Press. Available online at <www.nap.edu/catalog/10353.html>.

Neufville, R. 2000. Prospects for the future. Proceedings of the Airports in the 21st Century Conference, April 20, Washington, D.C. Transportation Research Circular Number E-C027. J.M. Rakas, G.W. Blomme, G. Gosling, eds. March 2001 (revised April 2001). Washington, D.C.: Transportation Research Board. pp. 90–113. Available online at <http://gulliver.trb.org/publications/circulars/ec027/ec027.pdf>.

RTCA Free Flight Steering Committee. 2002. National Airspace System Concept of Operations and Vision for the Future of Aviation. Washington, D.C.: RTCA, Inc.

2

Improving the Air Transportation System

The air transportation system is changing and will continue to change. Over the long term, however, it will be difficult for the air transportation system to change rapidly enough to meet changing requirements related to capacity, environmental effects, consumer satisfaction, safety, and security, while meeting ongoing requirements for the economic viability of service providers.

Most efforts to increase system capacity are focused on evolutionary or incremental changes that address specific constraints while aircraft are en route, in terminal areas, or on the ground at airports. For example, parallel arrival streams may be used when airport visibility is restricted to restore capacity to levels typical of clear weather (VMC, or visual meteorological conditions). Another option would be the use of advanced systems to adapt traffic flow in response to convective weather fronts to minimize or eliminate reductions in capacity. Meeting demand over the next 25 to 50 years, however, is likely to require a more revolutionary approach that seeks to increase capacity significantly beyond the level that the system currently enjoys even under ideal weather conditions. This may require completely different system operating concepts. The key point is that long-term goals may need to focus on fundamental, revolutionary structural changes in the air transportation system. One approach to defining future operational concepts is to propose solutions to shortcomings in the current system. To facilitate revolutionary change, however, a better approach would be to begin with a vision of the capabilities desired for the air transportation system of the future and then investigate how to provide those capabilities.

IMPETUS FOR CHANGE

The air transportation system in the United States and around the world is changing and will continue to change in response to many different factors. Although new technologies have the potential to enable more efficient operations, the current economic crisis faced by most major airlines makes it difficult for them to make large investments in new technologies or infrastructure. Public concerns about safety and security, poor conditions in the general economy, and other factors that temporarily suppress demand can be economically devastating to the air transportation industry. Demand for air transportation services can also be suppressed by the reluctance of passengers to travel when delays from undercapacity or intrusive security procedures become too onerous. Airlines' business decisions are also constrained by the competitiveness of the industry and the desire of each service provider to maintain market share, regardless of the economic climate. However, temporary setbacks notwithstanding, demand for passenger and cargo services has always increased over the long term and is expected to continue increasing in the future.

Even before 9/11, the government played a key role in ensuring safety, in part because it is so difficult for the public to assess risks in systems as complex as the air transportation system. In addition, high-profile accidents and incidents create tremendous public and political pressure for the government to act, even when it's too early to know for sure the cause of a particular tragedy. Other factors that will shape the future of air transportation include new methods of communications and other changes in the world that may cause the demand for business travel, leisure travel, or air cargo to grow more slowly—or faster—than current long-term projections.

Predicting the future of the air transportation system is difficult because it depends on the actions of—and interactions among—many different factors and organizations, many of which are themselves changing in ways that are unprecedented and hard to predict. Nonetheless, the parameters used to measure the performance of the system—comfort, convenience, costs, and societal impact—are not likely to change any time soon, and they can be used to guide the development of technologies even if the environment in which they will be employed cannot be precisely determined in advance.

OPERATIONAL CONCEPTS

Operational concepts can be used to describe how the air transportation system might advance, from the reasonable certainty of near-term requirements, technologies, and schedule implementation to a less certain vision of the long-term future.

Today there is no single national vision for the air transportation system 25 to 50 years from now. The visions that do exist, however, have a unifying theme—namely, improving performance in terms of capacity, environmental effects, safety, and security. Chapter 1 describes a larger set of system performance parameters that future visions should embrace. Existing public policy on access to the airspace and equitable use of the facilities in the air transportation system is expected to continue, and air operations are expected to increase overall, growing with the population and the economy. Long-term operational concepts should support a broad vision that encompasses all of these expectations.

Near-term operational concepts should ideally be derived from clearly understood transportation system needs. The pace of their implementation will be limited by the availability of mature technology and a host of nontechnological factors. Long-term operational concepts can serve as a guide for examining technological and nontechnological proposals and societal presumptions. To prepare for the future, a range of operational concepts should be developed, examined, and revised using an iterative process that considers potential changes in technology, society, and the air transportation system itself. This requires the ability to test and examine operational concepts for the future in a comprehensive manner. For example, these operational concepts should consider environmental needs and benefits. In the future it may be desirable to control the cruise altitude or flight path of an aircraft to avoid the formation of contrails that affect climate. In general, improvements in system efficiency can be expected to improve environmental performance by reducing fuel consumption, but trade-offs between emissions and community noise may need to be balanced.

The process of developing operational concepts also provides an opportunity to achieve national consensus among the various agencies and stakeholders at a level of detail that permits more focused agreement and planning. The salutary effect of this unifying activity is that it can stimulate and guide research in both technical and nontechnical areas.

The FAA's Operational Evolution Plan represents a general consensus on one way to bring known technology, infrastructure development, and system needs together and implement them to increase the capacity of the air transportation system over the next 5 to 10 years. However, the Operational Evolution Plan is not intended as a basis for examination and testing of longer term concepts or as a guide for research that will impact needs over the next 25 to 50 years. In particular the modeling and simulation tools of today are not sufficient to evaluate many long-term concepts and transition issues.

The vision published by RTCA, Inc., discusses how to accommodate growth in demand through 2020 and beyond.[1] It states that "operations are increasingly aircraft centric, focusing on performance rather than equipment standards, with use of required navigation performance as a key step in enabling greater efficiency, flexibility, and capability enhancements. Access to real-time information for decision-making supports efficient operation of the air transportation system when capacity limitations such as weather adversely impact the system. Enhanced system supported coordination and decision support capabilities allow the system to migrate beyond human centric operations" (RTCA Free Flight Steering Committee, 2002). The same RTCA document contains an evolutionary concept of operations that proposes changes in the air transportation system in three time periods (through 2005, 2005 to 2010, and beyond 2010). As with any future operational concept, this concept should be tested through simulation and modeling to estimate the technological and nontech-nological needs, benefits, and costs. The concept should then be refined and reevaluated as a basis for guiding research, identifying transitional issues, and determining if it is likely to succeed as a unifying effort in guiding future development of the air transportation system.

Looking out to 2050, it is not too early to begin identifying notional operational concepts, developing evaluation tools, and supporting research that enables the process to go forward. Simulation and modeling capabilities more powerful than today's will be required to better understand the complexities of the suite of systems that comprise or will contribute to the future air transportation system. An iterative operational planning process is essential for articulating the direction in which the air transportation system is most likely to proceed as performance improves.

Finding 2-1. The Challenge. Developing meaningful and useful operational concepts stemming from a broadly defined vision of the air transportation system 25 to 50 years hence is a critically important task in the process of improving the performance of the system.

Recommendation 2-1. Operational Concepts 2050. The federal government, working with other stakeholders in the air transportation system, should develop a coherent set of

[1]RTCA, Inc., is a not-for-profit corporation that functions as a federal advisory committee in developing consensus-based recommendations on contemporary aviation issues in support of the FAA and other elements of the aviation community.

operational concepts to support a vision for the air transportation system in 2050 to guide (1) long-term research and (2) the evolution of and transition to a more advanced air traffic management system. The set of operational concepts should be continually, objectively, and rigorously evaluated (for example, through comprehensive simulation and modeling) and iterated to reflect feedback from stakeholders, conflicts between alternative concepts, and the best understanding of the future costs, benefits, and requirements that are likely to evolve in response to changes in the real world, the current state of technology and systems operations, and future expectations. Strong national leadership should coordinate the efforts of all involved federal agencies and other stakeholders in the air transportation system to build toward concepts that best support the vision.

RESEARCH AND TECHNOLOGY

To a large extent, operational concepts dictate specific technology needs. However, regardless of the specific operational concept, many attributes of the future air transportation system can be predicted that point to general research and technology needs.

The first attribute is that the future air transportation system will involve much more automation both on the ground and in the air. Many modern aircraft are already so highly automated that, once programmed by the pilots, they can perform almost all guidance, navigation, and control tasks autonomously. This automated capability would need to be enhanced, however, to fit into many future operational concepts that require new functions—for example, required time of arrival at fixes, self-spacing or station-keeping, and self-separation. The modern air traffic control and management system is not highly automated, and it may prove nearly impossible to develop and test the underlying algorithms for fully automatic control in all situations, especially in the face of disruptions and emergencies; the same is generally true for airline operations centers. Therefore, some functions may be fully automated (e.g., aircraft guidance), others may be supported via automated decision aids (e.g., controller decision aids; and automated monitoring and alerting systems), and still others may rely on human decision making while using information systems for communications, visualization and situation assessment, and prediction of future conditions. The automation of many of these functions requires continued research and development.

Second, humans will be an integral part of the future air transportation system until the (unforeseen) day when the system can be automated to the extent that it requires neither intervention nor monitoring. Rather than framing the allocation of functions as a matter of "machines versus humans," emphasis should be placed on creating synergy between humans and machines where their combined performance is better than either alone could achieve. Substantial research into flight deck automation has demonstrated a wide range of problematic interactions between humans and automation that can be generalized to broader applications in air traffic management; other studies demonstrate similar issues with decision aids (Wiener and Curry, 1980; Sarter and Woods, 1992; Layton, Smith, and McCoy, 1994; Pritchett, 2001). Automation design often appears to be driven by technological capability with neither (1) sufficient insight into its functioning within the larger system nor (2) the ability to predict commensurate changes in coordination between system elements and the training required of human operators. Automation must be demonstrated to work with humans in the larger context of system performance in both nominal and off-nominal conditions. Additionally, the humans in the system will also require coherent procedures and training designed in concert with the technology.

Third, the future air transportation system will be more fully integrated. For example, systemwide optimization of traffic flows may negate the effectiveness of localized traffic flow management within air traffic control centers and sectors unless it is integrated into a nationwide discussion at all levels of air traffic operations. Likewise, functions traditionally assigned only to aircraft, sector controllers, traffic managers, or industry representatives (e.g., airline dispatchers) will need to incorporate joint decision making that involves several entities and considers their disparate objectives.

Fourth, the integration of functions into the future air transportation system will require distributing responsibility and decision making between and among disparate entities. These entities may be geographically distributed and will often represent the interests and viewpoints of different organizations. The distribution of authority and responsibility with a large system presents technical and organizational challenges that should be studied and evaluated rigorously. Likewise, enabling distributed operations will require more insight into communication, coordination, and collaborative work mediated over distances via information technology and automation.

Fifth, the future air transportation system will be complex by almost any measure of complexity, yet will need to achieve the highest levels of performance and safety in a wide range of anticipated and unanticipated conditions. The ability to imagine changes to the system outpaces the ability to develop implementing technologies and procedures, integrate them into a reliable and highly capable air transportation system, continuously operate the system, collect and assess data on system performance, and make future improvements. In addition to the simulation and modeling capabilities recommended in Chapter 3, suitable system engineering models are needed for guiding systems analysis, design, integration, and implementation, especially in the case of large software developments.

Finally, aircraft separation standards, at least where they form bottlenecks that limit system capacity, will need to be reduced. Current separation standards were based on system shortcomings that future technologies may address. Some of these factors are related to aircraft design and are described in Chapter 4. Factors relevant to air traffic management technologies include the following:

- Errors in control and knowledge of aircraft position, which might be reduced or functionally eliminated by ubiquitous and transparent communication, navigation, and surveillance technologies.
- Lack of situation awareness, especially with regard to current and future separation, which might be mitigated by improved sensors and displays, such as synthetic vision, cockpit display of traffic information, and controller displays.
- Safety buffers to account for monitoring failures and late detection of potential conflicts, the size of which might be reduced by air- and ground-based conflict detection and resolution systems.
- Wake vortices, which might be better understood and predicted or which might be sensed and avoided in real time.

Advanced technologies in some of the above areas could also produce important secondary benefits. For example, technology to directly sense the magnitude and location of wake vortices might also help avoid clear air turbulence, which is an ongoing threat to safety and passenger comfort.

Recommendation 2-2. Enabling Technologies. Enabling technologies applicable to a wide range of operational concepts should be developed in parallel with development and evaluation of long-term operational concepts so that the necessary technologies will be ready for whichever operational concept proves to be most beneficial. Technology areas of particular interest include the following:[2]

- Automation technologies applicable to fully automated systems; automated decision aids; and information systems for communication, visualization, situation assessment, and the prediction of future conditions.
- Technologies that support distributed, collaborative decision making and that foster coordination and interactions among multiple human and automated elements of the system.
- Methods and technologies for moderating and abating the impact of noise and emissions locally, regionally, and globally.
- Methods and technologies for predicting or directly sensing the magnitude, duration, and location of wake vortices and the potential to reduce separation standards without compromising safety.
- Methods for identifying (1) the information required for situation awareness when humans are assigned novel (untried) tasks in future operational concepts and (2) sensor, computing, and display technologies for better supporting situation awareness, judgment, decision making, and planning. Relevant technologies include synthetic vision, cockpit and controller displays for novel air traffic management functions, fast-time simulation and computational functions for predicting future conditions, and alerting. These methods and technologies should be investigated for their potential to (1) reduce separation standards without compromising safety and (2) enable changes in the roles of humans within the system.
- Systems-engineering methods that are (1) capable of conceiving and analyzing systems of the complexity of air transportation and (2) suitable for governing the design, testing, and implementation of these systems.
- Avionics technologies that will provide ubiquitous and transparent communication, navigation, and surveillance capabilities; enable cost-effective, reliable air traffic management; and contribute to the reduction of separation standards without compromising safety.

Recommendation 2-3. Design of Complex Human-Integrated Systems. The design of human-integrated systems—that is, systems that rely on the combined activities of humans and machines—presents significant challenges at every level, from the systems level (e.g., creating effective teamwork within operations involving many human operators and automated system elements) to the detailed design level (e.g., developing operating procedures and system displays). Research in the following areas is required to understand and address these challenges:

- A broad, interdisciplinary approach that includes technology designers, users, and experts in human and organizational performance from the earliest stages of conceptual design through final implementation to develop technology that effectively supports human behavior and recognizes the need for concurrent design of procedures, training, and technology.
- Geographically distributed activities, such as coordinated decision making and planning, that are mediated by computers and automated system elements.
- Human factors, human-automation interactions, and functioning of teams of humans and automated system elements.
- Specific impact of newly automated functions and changes in human roles.
- System engineering methods for addressing organizational and systemwide issues.

[2]In this and other recommendations that list research areas, the bulleted items are either listed alphabetically or grouped topically. The committee did not prioritize research areas in each list.

BEYOND TECHNOLOGY DEVELOPMENT

The air transportation system includes aircraft, air traffic control and air traffic management systems covering every phase of flight, airports, labor, airlines, and other organizations involved in research, development, manufacture, operation, certification, and regulation of aircraft and aviation systems. The previous sections of this chapter focus on air traffic control and air traffic management systems. Chapter 4 focuses on aircraft and aircraft technologies. These are the segments of the air transportation system where government research and technology development have the most direct impact. However, the ability to introduce and manage change, including technological change, is also a function of many other factors. The federal government, in particular, has tremendous leverage in its power to set economic policy, regulate the aviation industry, and collect and disburse billions of dollars in aviation taxes and general revenue each year. In addition to technological research to improve the performance of aircraft and air traffic management systems, the air transportation system would also benefit from research that addresses institutional issues; processes for modifying regulations, certification requirements, and operating procedures; societal concerns about aircraft noise and emissions; demand; and economic factors.

Most organizations fear both technological and business risk as well as changes that could create risk. Although current organizational structures and policies have shortcomings, they tend to be known and manageable. Change offers the potential to improve the current situation, but it also creates uncertainty and the risk of unforeseen consequences. Change is of particular concern if it could damage the vested interests of some organizations (e.g., by changing existing job descriptions or organizational missions or, in the extreme, by eliminating jobs or business units). Change will also be resisted if it might allow some organizations to succeed at the expense of others. All of this creates tremendous inertia that must be overcome to change the status quo. Along with strong leadership (see Chapter 1), the air transportation system would benefit from research on processes to predict, identify, and resolve the conflicting objectives of different stakeholders. Such a program of research should recognize that air passengers, shippers, and aircraft owners pay the bills of the other stakeholders, even though customers often are not directly represented in stakeholder debates about the future vision. With the ultimate customers kept in mind, it is still possible, however, to suggest specific research to avoid or minimize the consequences of behavior that undermines the overall effort to implement new operational concepts and achieve the future vision.

The FAA must certify new aircraft and air traffic management systems and approve operational procedures prior to use. Current handbooks used in the certification of aircraft and aircraft systems do not cover many innovative system concepts, such as the shift of some air traffic management responsibilities to the cockpit. For such systems, criteria for certification and operational approval will need to be developed concurrently with the systems and procedures themselves to prevent substantial delays in implementation. Improved processes are also needed to (1) facilitate changes to current operational concepts and (2) implement new operational concepts and the new technologies needed to support them. In many cases, proposed changes will need to be coordinated with other nations and international organizations prior to implementation. Aircraft manufacturers and airlines, responding to the changing market for their products and services (as well as new government policies), make choices determining the size, speed, fuel efficiency, environmental characteristics, and other performance parameters of new aircraft. Those choices influence the comfort, cost, convenience, and societal impact of air transportation and hence the aggregate level of commercial air transportation activity. The structure of the airline industry and the operational strategies followed by the individual airlines evolve in response to government decisions and policies in three broad categories:

- Public and private research and development efforts that produce the particular facilities, equipment, and systems available to manufacturers and airlines.
- The provision of infrastructure and support services, principally airports and air traffic control services, and the related system of taxes and fees imposed at the national, state, and local levels.
- Rules and regulations established by U.S. and foreign governments and by international regulatory bodies regarding operational procedures, safety, and business practices.

Economic factors directly affect system demand and capacity and levels of service available to various system users. Airlines attempt to maximize economic performance through decisions that weigh the impact of incentives and penalties built into the system of rules, regulations, taxes, and fees. These incentives and penalties should be carefully constructed to avoid encouraging behavior that makes it more difficult to achieve the future vision. Currently this may not be the case: The cost-benefit analyses used to justify new certification standards and regulations often lack credibility with the owners, operators, and manufacturers who must bear the costs of implementation. In addition, the process for setting U.S. government rules, regulations, taxes, and fees for airfield and airways capacity is not well supported by economic research that considers the likely responses of system operators and users to changes in aviation economic policies. For example, the tax on airline passenger tickets is calculated at $3 per passenger enplanement plus 7.5 percent of the value of the ticket. Because of the wide range in ticket prices, passengers on the same flight receiving the same level of service will be assessed different levels of tax, even though passengers paying higher fares impose no more burden on the air transportation system than do discount passengers on the same flight. Weight-based landing fees exacerbate the distortions of the ticket tax. Large aircraft carrying many passengers impose essentially the same burden on system capacity as smaller aircraft. Large aircraft may require a larger investment in runways, but not in proportion to the higher fees they must pay. In fact, a small aircraft may place a larger burden on the air traffic management system if it has a low approach speed and must be merged into a landing stream of large aircraft with higher approach speeds.

The size, speed, fuel efficiency, environmental characteristics, and passenger comfort offered by future generations of aircraft, as well as the capabilities of the air traffic management system, will directly influence the cost and convenience of commercial air transportation and, hence, the aggregate level of demand for air transportation services. In order to appreciate the costs and benefits, understanding economic factors is especially important in small communities where the government subsidizes commercial air service because it cannot be justified based purely on market factors. Economic analyses should also be used to help assess different approaches for improving capacity—for example, by assessing the feasibility of various economic incentives or by comparing the cost of building more runways with the cost of developing a more capable air traffic management system that increases the capacity of existing runways. Improving safety and reducing environmental effects can reduce costs in terms of total, long-term costs and even, in many cases, of direct operating costs. Foresight, planning, and vision play an important role in determining the feasibility of achieving future goals; costs and consequences need to be recognized early on rather than waiting until after a system is deployed to recognize, for example, that it creates noise or air quality problems that will limit its implementation and benefits.

Finding 2-2. Nontechnological Impediments to Success. Technological research alone is insufficient to achieve the future vision. Research is also needed to (1) better understand the economic, environmental, political, institutional, and managerial factors involved in achieving key goals, (2) take advantage of synergies among these factors, and (3) overcome related impediments.

Recommendation 2-4. Research Needs Beyond Technology Development. The federal government should also support research to develop improved processes and methods in the following nontechnology areas:

- Assessment of economic factors, such as taxes, fees, and subsidies established by the government, that influence (1) the demand for and the supply of air transportation services and (2) key decisions made by organizations and individuals involved in the provision and use of the air transportation system.
- Modification of regulations, certification requirements, and operating procedures.
- Prediction and resolution of conflicting objectives of different stakeholders in the air transportation system.
- Understanding societal concerns about aircraft noise and emissions.

REFERENCES

Layton, C., P. Smith, and C. McCoy. 1994. Design of a cooperative problem-solving system for en route flight planning: An empirical evaluation. Human Factors 36(1):94–119.

Pritchett, A. 2001. Reviewing the roles of cockpit alerting systems. Human Factors in Aerospace Safety 1(1):5–38.

RTCA Free Flight Steering Committee. 2002. National Airspace System Concept of Operations and Vision for the Future of Aviation. Washington, D.C.: RTCA, Inc., p. v.

Sarter, N., and D. Woods. 1992. Pilot interaction with cockpit automation: Operational experiences with the flight management system. International Journal of Aviation Psychology 2(4):303–321.

Wiener, E., and R. Curry. 1980. Flight-deck automation: Promises and problems. Ergonomics 23(10):995–1011.

3

System Modeling and Simulation

The committee believes that modeling of the air transportation system is best accomplished by a suite of system models.[1] This chapter describes the committee's understanding of system models and how they can be used to design and analyze evolutionary and revolutionary operational concepts, technologies, and other changes to the air transportation system. Given existing modeling and simulation capabilities (and ongoing research), the chapter also suggests what else should be done, especially by government, to provide the long-term systems modeling capability needed to analyze and select changes to the air transportation system.

UNDERSTANDING SYSTEM MODELS

In its simplest description, a suite of system models is a set of models, each self-contained and designed to produce meaningful outputs by itself, where outputs from some models are used as inputs to other models. As a general rule, the suite includes very detailed, high-fidelity, data-intensive, long-run-time models, usually involving individual components of the overall system, as well as higher-level, fast-time, abstract analytic models that, at the highest level, seek to represent how the entire U.S. air transportation system functions and how that functioning impacts the economic vitality of the nation.

Fundamental to the use of a suite of system models is the recognition that it is not possible to capture all of the important variables within a single large model, nor do the models in the suite have to be directly connected or operate simultaneously. Moreover, and of equal importance, outputs from the more detailed models often provide insights into the causes of congestion in the air transportation system and the relevance of potential actions aimed at addressing the problem. If these outputs were simply passed along to a high-level analytic model, such insights might well be lost. A suite of systems models should include a variety of models, some simpler, cheaper, easier, and quicker to run (when they can provide the needed output results with the required level of accuracy) and others more complex, more expensive, more difficult, and slower to run (when more detailed and/or more accurate results are needed and worth the extra effort and expense).

Computer-based simulations range from large-scale, fast-time simulations of the entire U.S. air transportation system to detailed human-in-the-loop simulations of specific aircraft or air traffic management systems. Simulations complement other analytical efforts by helping to (1) determine the feasibility of operational concepts, (2) establish parametric values required by models (for example, the increase in airport capacity that a new operational concept or technology will produce), and (3) validate model assumptions.

Also of importance is the strong interdependency of the many factors that enter into assessments of the air transportation system (see Figure 3-1). Thus, when constructing a suite of models, it is critical to capture logical dependencies and interdependencies and to make sure that the available models accurately simulate each of the areas depicted in Figure 3-1 or that efforts are under way to develop better models.

Detailed models support decisions on improving individual elements of the air transportation system. A suite of system models should be designed to assess the performance of system elements and the system as a whole. Incompatibilities that limit the ability of detailed models to support broader analyses should be avoided. High-level abstract models cannot include many of the variables that are in the more detailed models. There is, therefore, a difficulty associated with mapping the sensitivities of the results of the

[1]The Department of Defense and some other organizations use the term "system of system models" for what this report calls a "suite of system models."

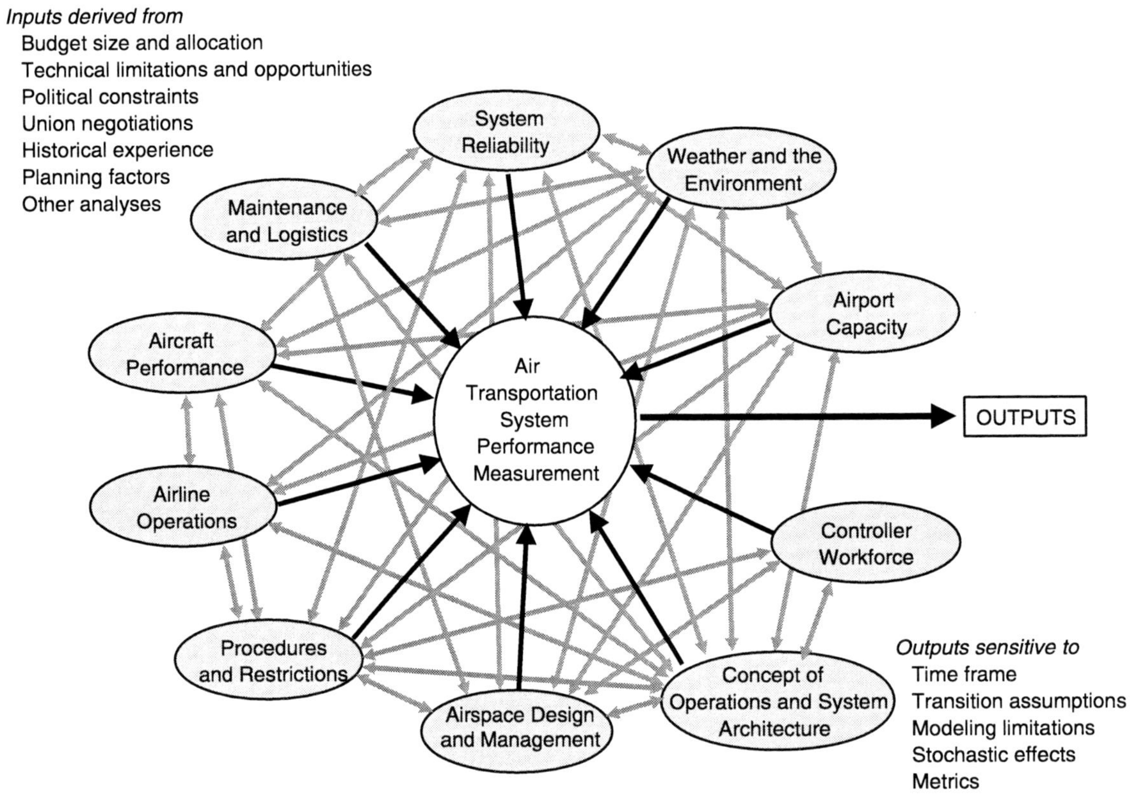

FIGURE 3-1 Generic inputs into an air transportation system performance model. Of particular note is the interconnectivity among the inputs, suggesting the need for substantial analyses at the input component level to understand sensitivities.

more detailed models into the higher-level models, so that higher-level models often cannot reflect such sensitivities, even though the more detailed models may show them to be important for problem identification and resolution. When reporting the complete results of a suite of system models, it is important to include the results of the more detailed models where they are relevant to the solution. A description of four levels of models that could be included in a suite of system models appears in Appendix E.

A particular challenge in using a suite of system models for a sociotechnical system as complex as air transportation will be to capture the nonlinear dynamics of interactions among components, which makes it difficult to combine the results from different models. Additional research is needed to overcome this challenge.

ANALYSIS AND DESIGN TO IMPROVE AIR TRANSPORTATION SYSTEM PERFORMANCE

Improving the performance of the air transportation system requires a good understanding of the operation of the current system and the ability to model and analyze the performance of new operational concepts. The air transportation system, however, is a complex, human-centered system that involves multiple technologies, organizational structures, human behaviors, and competing economic entities. Modeling such a complex system is extremely difficult and requires the ability to model interdisciplinary systems and operational concepts (including cross-functional operational concepts) in terms of system performance (comfort, convenience, costs, and societal impacts) and the ability to satisfy the often-conflicting objectives of various stakeholders. Improving the ability to model and measure systemwide performance and assess risks associated with the development, deployment, and operation of complex new systems will help avoid historical precedents in which large new system projects have been cancelled prior to completion because of delays, cost increases, and/or the inability to meet design requirements.

As system complexity increases, it becomes more difficult to guard against dysfunctional interactions. Problems may arise from unanticipated interactions among automated subsystems and from unanticipated interactions among different organizations and parts of organizations. The ability to develop complex systems while effectively managing problems at the intersections between organizations, disciplines, and systems is growing more slowly than the ambition and willingness to attempt the development of such systems. Interdisciplinary research and a systems approach to research are needed. Business as usual, with research segregated by discipline, is insufficient and runs the risk of (1) optimizing short-term performance at the expense of long-term improvements or (2) suboptimizing system performance (i.e., optimizing the performance of a portion of the system in a way that fails to optimize or even degrades total system performance).

Demand Models

Demand and demand allocation do not remain static in the presence of changes to the air transportation system and the world. Passengers, airlines, manufacturers, business aviation, general aviation, and other involved parties will all adjust their behavior in response to increasing capacity and other changes. Models should be able to account for these changes. Overly simplistic modeling in the past missed the surge in demand following deregulation and the advent of hub-and-spoke operations, the emergence of low-price point-to-point carriers, and the rise in regional jets. Models should also be able to account for changes in the behavior of individual airports and regional airport systems, including the construction of new runways, gates, and other facilities. Models like the Total Airspace and Airport Modeler (TAAM) would be highly useful for evaluating some of the above factors, but additional models are also needed. Developing models capable of learning and adapting (e.g., agent-based modeling) is therefore very important. Models that account for mode splits (e.g., competition among air, rail, and automobile travel) in selected corridors will also play an increasingly important role.

In the specific case of the U.S. air transportation system, the above improvements are needed to produce suites of system models to do the following:

- Characterize the nature of future demand as a function of possible changes to the price, quality, and availability of complementary and competitive transportation services; the overall performance and structure of the air transportation system; perceptions of aviation security; the personal habits and tastes of consumers; consumer income; and other factors internal and external to the air transportation system.
- Identify potential shortfalls and needs in the performance of the U.S. air transportation system due to future growth in demand.
- Determine the ability of new technologies, operational concepts, and procedures to meet future shortfalls.
- Assess the systemwide aviation impacts of adapting evolutionary and/or revolutionary technologies and operational concepts and determine the overall benefits and costs of various alternatives.

Evolutionary and Revolutionary Approaches

The requirements for and capabilities of technologies, procedures, and operational concepts may be analyzed using two different approaches. The evolutionary approach starts with the operational concept and technologies used in the current air transportation system and determines the impact of incremental changes to them (see Figure 3-2, left-hand side). These changes are understood using technology models, computational human performance models, and human-in-the-loop simulations using increasing parametric abstraction and emulation. Impacts on the operation of the overall air transportation system are then determined in terms of metrics such as delays and flight times, using current demand and projected future demand. The benefits and costs are then evaluated for each of the evolutionary improvements.

In the revolutionary approach, the analysis starts with the top-level functional and performance requirements that are necessary for the system to meet various levels of future demand (see Figure 3-2, right-hand side). This approach may be viewed as revolutionary in the sense that totally new operational concepts, architectural approaches, system characteristics, and technological capabilities may be postulated without first assessing their feasibility or relationship to the existing system. Once the system is defined at the top level, parametrically connected layers of models may be used to allocate functional and performance requirements to system elements and human operators. Trade-off studies of alternative concepts and postulated technological capabilities can then be analyzed in terms of benefits, costs, and risks.

The two approaches differ in terms of the starting point of their analyses, not in the nature of the models used to implement them. In both approaches, system analyses must be iterated to account for interactions among various factors—internal and external to the air transportation system—that affect system performance and demand. Given that the functional description of the air transportation system can be likened in a simplistic way to a network of capacity-constrained links and nodes, it is clear that the detail and fidelity of models used in both the evolutionary and revolutionary approaches need to be similar at corresponding levels. In the revolutionary approach, the links and nodes should have specific capacities to meet potential levels of future demand (that is, capacities are assigned to the links and nodes in a way that achieves the desired level of overall network flow performance). New operational concepts and technologies are evaluated to see if they can achieve the requirement by using more detailed technology models, human performance models, and, ultimately, human-in-the-loop simulations. In the evolutionary approach, we modify existing operational concepts and technologies and then evaluate the extent to which the changes accommodate future demand using fast-time network flow models.

The approach used to analyze human behavior differs as

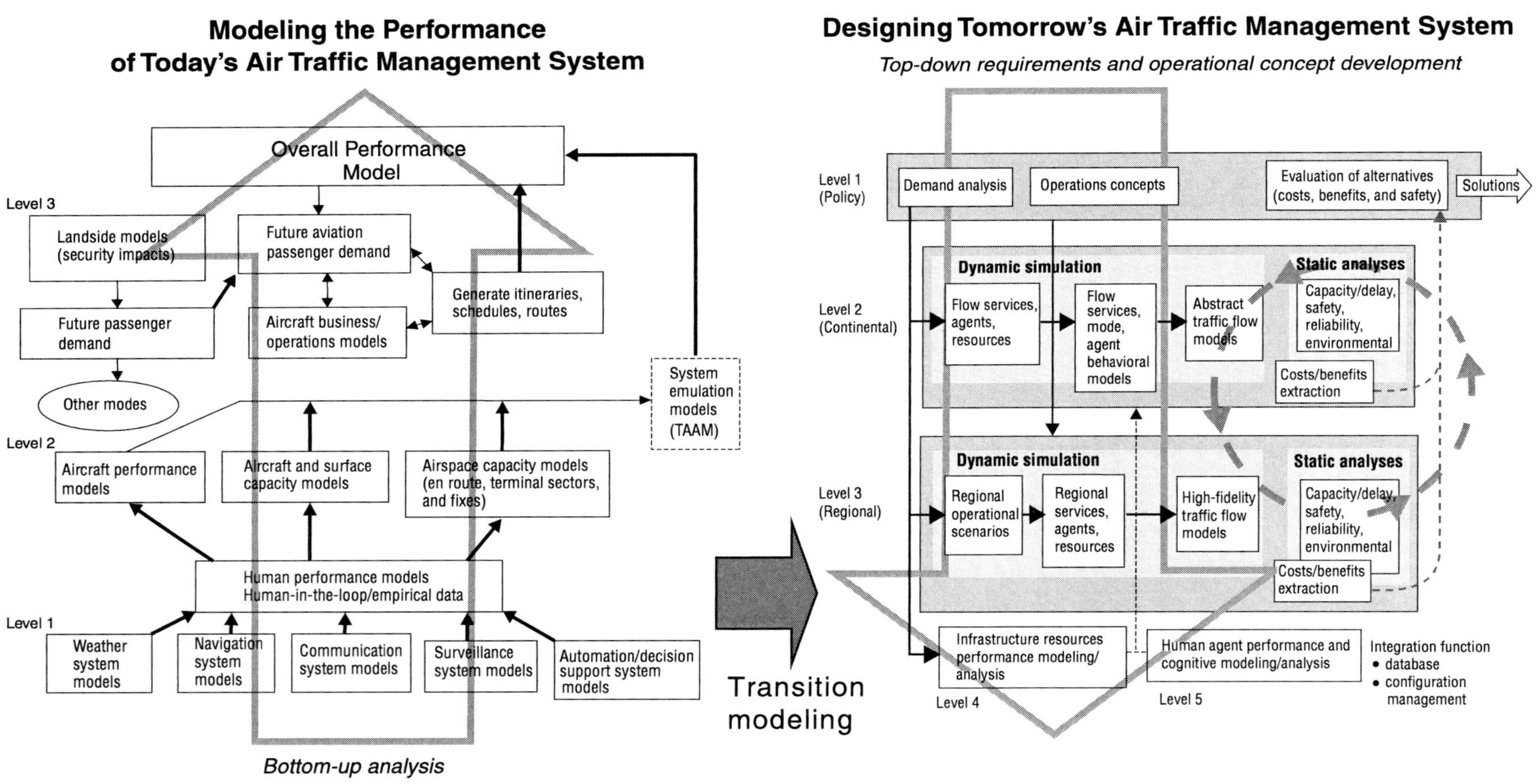

FIGURE 3-2 Fundamental air traffic management modernization requires analytical approaches with two different starting points. Source: Dennis Muilenburg, Boeing, briefing to the committee, November 26, 2002.

well in the evolutionary and revolutionary approaches. The evolutionary approach is baselined in current operations, where controllers, traffic flow managers, pilots, and dispatchers all have set roles and procedures. Proposed changes in technology (both hardware and software), roles, and procedures are then tested and refined using human-in-the-loop simulations, and performance improvement is measured. In the revolutionary approach, performance parameters and human roles are allocated to satisfy top-down functional requirements. Progressively more detailed simulations using human performance models can then be used to determine feasibility; ultimately, human-in-the-loop simulations can verify predictions, but their development requires detailed design of interfaces and operating procedures as well as the training of personnel on the revolutionized operational concept, operating procedures, and new technologies.

There exist or are under development today in industry and government a number of models and human-in-the-loop simulations that can fill roles at various levels in the overall evolutionary and revolutionary suite of system model constructs described above. These include, for example, the FAA Technical Center's Integration and Interoperability Facility, NASA-Ames's Virtual Airspace Modeling and Simulation Project, the approach used by Boeing in developing its Discrete-Event Simulation Interactive Development Environment (DESIDE), MITRE's Detailed Policy Assessment Tool (DPAT) model, and the Logistics Management Institute network simulation model (LMI Net). Professional and expert-to-expert consultations among these efforts exist. What is missing is a uniform federal strategy for research investments, interagency coordination, and the use of the modeling and simulation results.

Modeling Gap

There is a significant difference in the detail, run times, and data requirements for the various models. On the one hand, models such as TAAM and facilities such as MITRE's real-time air traffic management infrastructure laboratory provide very detailed emulations of the U.S. air transportation system at the expense of long run times and extensive data preparation. DPAT and LMI Net, on the other hand, are fast-time models that permit high-level evaluations of air transportation system performance by sacrificing the ability to produce detailed intermediate data. Research to develop improved intermediate models that close this gap could be of considerable benefit.

Validation

The stochastic nature of the air transportation system ensures that no one model gives a precise answer. Results on days that are supposedly equivalent from a scheduled airline viewpoint can differ wildly for a variety of reasons:

- Changes in the sequence of actual takeoffs for a number of aircraft scheduled to depart within the same 15-minute window.
- Changes in the number of military, business, and general aviation operations.
- Changes in wind direction at one or more major airports that require a change in airport configuration.

As a result, key questions remain unanswered: What does it mean to validate a suite of system models? How should the validation be conducted? Who—that is, which entity—should certify the degree of validation?

When linking established models, the validation challenge becomes more significant. For example, connecting two validated models (in run time, or by using the output of one as input to the other) does not guarantee that their combined output is itself valid. In other words, establishing mechanisms for combining models is itself a modeling process that must detect and account for conflicts and gaps that may exist among the assumptions and capabilities of each component model.

Answering the above questions and developing widely accepted validation standards and processes will not be easy, even for an organization with the resources of the federal government. Areas of particular difficulty include predicting strategic investment decisions by industry and government, such as hub selection and location and airport construction projects, which (1) depend on a complex interplay of public and private individuals, organizations, and interests and (2) change the shape of the landscape upon which the rest of the air transportation system rests.

Federal research investments in a suite of system models relevant to the air transportation system need to be better coordinated to avoid unnecessary gaps and overlaps. Widely accepted criteria are needed to validate new models and updates to existing models, and a library of validated models is necessary to moderate user contributions to the models and, ultimately, to support good policy decisions by users inside and outside government. The history of NASTRAN, the NASA Structural Analysis program, provides a good example of the government developing an important software tool and then making it widely available to users and to commercial developers, who used it as the basis for creating additional applications and analysis tools. NASA initiated the development of NASTRAN in the early 1960s to provide its aerospace research projects with a finite element analysis capability. The initial release of NASTRAN, in 1968, was primarily of interest to the large aerospace companies and government laboratories that could afford the multimillion-dollar computers necessary to run the software. Since then, improvements in the NASTRAN code by users have extended the applicability of NASTRAN to almost every kind of structure, and improvements in the capability of widely affordable computers have removed computational capability as a limiting factor in the use of NASTRAN. NASA's decision to make the NASTRAN source code available to users and developers also contributed to the tremendous expansion of NASTRAN's capabilities. Emulating the precedent of NASTRAN with a broad suite of air transportation system models would be difficult because of the resources required, the intellectual challenge involved, and the proprietary nature of many models, which are viewed by their developers as a means of maintaining a competitive edge with respect to other modeling organizations. Nevertheless, the benefits of such an effort, if successful, would be substantial.

SYSTEM MODELS AND AIR TRANSPORTATION SYSTEM SAFETY

Safety analysis of air traffic management operational concepts has traditionally been based on chain-of-event models, but other approaches based on systems theory have recently been proposed.

When safety is characterized by a chain of events, it may be analyzed using hazard analyses, in which the events leading to each hazard are identified (e.g., fault tree analysis or failure modes and effects criticality analysis) and deterministic models are built of the combinations of failure events and human errors. The hazard analysis models may be used to redesign the system such that hazards are eliminated or mitigated. In addition, probabilistic analysis of events and chains of events is sometimes used to determine the risk associated with a design. Various types of formal mathematical analysis can be applied to state-based models (both probabilistic and nonprobabilistic) to evaluate various aspects of safety.

In addition to using formal analysis to evaluate safety, simulations might be used. These simulations must include humans, who are an integral part of the air transportation system. One approach to the problem, proposed by, among others, Gore (2000) and Pritchett et al. (2001), is to use large-scale, agent-based simulations spanning one or more traffic sectors. Ultimately, it is hoped that these simulations will be able to simulate with high fidelity the ability of agents to reason and react to unexpected situations, but progress will depend on the ability to build accurate models of human behavior.

Simulation has also been proposed as a way to extend risk assessment methods. Such simulations build on traditional hazard analysis models but enable the use of nontraditional event ordering. By allowing for inconsistent or variable event ordering, which can have a significant impact on whether a set of events leads to an accident, these simulations can examine a larger range of potential chains of events. The underlying models used for these simulations are commonly state-based, but other types of models might be used. Methods using stochastic, state-based models in the simulations have been proposed to substantially reduce the simulation runs needed by classic Monte Carlo methods (Blom et al., 1998).

Alternatives to event-based models have been proposed, primarily based on concepts of systems theory (e.g., Rasmussen, 1997; Svedung and Rasmussen, 2002; Leveson et al., 2003). Systems theory is the mathematical foundation for system engineering, with roots that go back to the 1930s (Checkland, 1981). Systems theory emphasizes the manner in which organized systems (both human and nonhuman) function. It includes the principles, models, and laws necessary to understand complex interrelationships and interdependencies among linked components and subsystems within a system. Systems theory models include organizational and managerial factors that are often omitted from chain-of-event models. Safety models based on systems theory view accidents as arising from interactions among system components (Perrow, 1984), where the interactions may be nonlinear and involve multiple feedback loops. Systems theory models can be used to analyze software-related accidents, complex human decision making, and system adaptation or migration toward an accident over time and can handle dynamic or behavioral complexity in addition to static or structural complexity.

In a systems theory approach to modeling, systems are viewed as interrelated components that are kept in a state of dynamic equilibrium by feedback loops of information and control. A system is not treated as a static design, but as a dynamic process that is continually adapting to achieve its ends and to react to changes in itself and its environment. The original design must not only enforce appropriate constraints on behavior to ensure safe operation, but it must also continue to operate safely as changes and adaptations occur over time. Accidents then are treated as the result of flawed processes involving interactions among people, social and organizational structures, engineering activities, and physical system components. Systems theory approaches to modeling and analyzing safety are new, and it remains to be seen whether the resulting models will be more or less effective than the traditional chain-of-event models.

The FAA has established, maintains, and continues to improve a suite of models that are used for environmental impact studies and assessments of proposed regulatory or market-based measures to control noise or emissions. Community noise models estimate the number of people exposed to high noise levels at a single airport (the Integrated Noise Model) or globally (the Model for Assessing Global Exposure to Noise from Transport Aircraft, MAGENTA) and can reflect the effects of noise abatement procedures. The FAA's local and regional air quality model (Emissions and Dispersion Modeling System, EDMS) calculates total emissions around an airport based on the number and type of aircraft operations and estimates how emissions are dispersed. The global emissions model (System for Assessing Aviation's Global Emissions, SAGE) inventories global emissions, summing emissions from each flight as a function of flight altitude and location. All of these models could be used as part of a suite of system models to evaluate the environmental benefits and trade-offs of measures to improve the air transportation system in terms of capacity and other performance parameters.

CONCLUSIONS

Given existing modeling and simulation capabilities and the state of ongoing research, the actions defined in the following recommendations would provide the long-term systems modeling capability needed to design and analyze evolutionary and revolutionary operational concepts and other changes to the air transportation system.

Recommendation 3-1. Value of Modeling and Simulation. Federal agencies involved in modeling and simulation of the air transportation system should make complementary use of field tests, laboratory tests, modeling, analysis, and simulation to improve their ability to (1) measure systemwide behavior of the air transportation system, (2) assess the performance of proposed operational concepts, technologies, and other changes, and (3) make informed investment decisions that reduce the schedule, cost, and technical risk of system improvements.

Recommendation 3-2. Management of System Models. Federal agencies that support research in aviation system models should improve their coordination, especially with regard to the following:

- Ensuring that the federal investment for research and development in aviation models focuses on key issues, avoids unnecessary duplication, and encourages cooperation among developers.
- Encouraging participation of industry and academia in modeling and simulation research and development relevant to government needs.
- Establishing widely accepted criteria for the maintenance and validation of models.
- Identifying models that are most important to government policy decisions.
- Making those models more widely available to users inside and outside government.
- Ensuring that modeling and simulation results are used appropriately by decision makers involved in developing the future aviation system.

Recommendation 3-3. System Modeling Research. The government and other interested parties should support additional research in the following critical areas:

- Improving the interoperability of high-fidelity, detailed, data-intensive, long-run-time models of the U.S. air transportation system and the higher-level fast-time, abstract models necessary to analyze overall system performance under a variety of different assumptions so that both types of models can be brought to bear on relevant problems. (It may be feasible to develop models with adjustable resolution that can simplify variables for faster run time when those variables are critical to the analysis being performed.)
- Modeling and simulation methods suitable for safety analysis, which inherently require a detailed level of modeling that includes all the factors that contribute to safety, including human performance and sociotechnical aspects of the system. (Additional fundamental research and development is required before these methods can enter widespread use. New approaches should be pursued using systems theory as well as new nontraditional chain-of-event models.)
- Modeling demand and demand allocation for air transportation services, particularly as it relates to airline schedule changes, including city-pairs, routes (including altitudes and way points), time of day, and the establishment (or elimination) of hub airports. (Dynamic interactions between changing or radically new operational concepts and technologies and user behavior, as they relate to all modes of transportation and other factors external to the air transportation system, must be better understood to ensure the right problems are being addressed.)
- Requirements, methods, and standards for validating individual models and suites of models.
- Understanding how to connect models to form a suite of system models that includes nonlinear dynamic interactions and emergent properties.
- Understanding the role of humans in the aviation system of the future and how to communicate this understanding in a convincing and supportable way. (Including computational human performance models in current simulations and using human-in-the-loop simulations is critical.)

REFERENCES

Blom, H.A.P., G.J. Bakker, P.J.G. Blanker, J. Daams, M.H.C. Everdij, and M.B. Kompstra. 1988. Accident risk assessment for advanced air traffic management. Proceedings of the 2nd USA/Europe Air Traffic Management Research and Development Seminar, Orlando, Fla. Also in G. Donohue and A. Zellweger, eds. Air Transportation Systems Engineering. 2001. Progress in Astronautics and Aeronautics Series. Reston, Va.: American Institute of Aeronautics and Astronautics, pp. 463–480.

Checkland, P. 1981. Systems Thinking, Systems Practice. New York: John Wiley & Sons.

Gore, B. 2000. The study of distributed cognition in free flight: A human performance modelling tool structural comparison. Third Annual SAE International Conference and Exposition on Digital Human Modelling for Design and Engineering, Dearborn, Mich., June 6–8. Available online at <www.engr.sjsu.edu/hfe/hail/airmidas/SAE2KGPaper_2181.PDF>.

Leveson, N., M. Daouk, N. Dulac, and K. Marais. 2003. Applying STAMP [Systems Theory Accident Modelling and Process] to accident analysis. Workshop on the Investigation and Reporting of Incidents and Accidents. Williamsburg, Va., September 16–19. Proceedings to be published by NASA Langley Research Center, Hampton, Va. (see <http://shemesh.larc.nasa.gov/iria03/index.html>).

Perrow, C. 1984. Normal Accidents: Living with High-Risk Technologies. New York: Basic Books.

Pritchett, A., S. Lee, and D. Goldsman. 2001. Hybrid-system simulation for national airspace system safety analysis. Journal of Aircraft 38(5):835–841.

Rasmussen, J. 1997. Risk management in a dynamic society: A modeling problem. Safety Science 27(2/3):183–213.

Svedung, I., and J. Rasmussen. 2002. Graphic representation of accident scenarios: Mapping system structure and the causation of accidents. Safety Science 40(5):397–417.

4

Improving Aircraft Performance

INTRODUCTION

Commercial jet aircraft produced in the United States are highly competitive, but they are also the result of technology investments made a long time ago. Technologies used to support the launch of the Boeing 777, the most recent model of U.S. widebody aircraft, were developed over 20 years ago. The effects of a diminished or misdirected aeronautics research program will not be significant in the near term, but eventually the result will be a diminished U.S. aeronautics industry.

Improvements in aircraft performance are critical to achieving necessary improvements in almost every aspect of the overall performance of the air transportation system (see Chapter 1). For airlines, operational cost (i.e., cost per seat-mile) is a key measure of aircraft performance. However, estimating the ultimate effect that long-term research and advanced technology may have on operational cost is often difficult at best.

A basic measure of aircraft productivity can be computed by multiplying payload by block speed (the average gate-to-gate speed for a given mission leg). Design efficiency is then indicated by the ratio of productivity to maximum takeoff weight (MTOW); high design efficiency is reflected in lower MTOW. A more complete understanding of aircraft productivity and efficiency should include additional factors, such as availability (the average number of hours per day, week, etc., that an aircraft can be operated, taking into account servicing and maintenance requirements), utilization (the actual number of hours per day, week, etc., that a particular aircraft is operated), operational range (which ideally should be matched to the routes on which a particular aircraft is employed), and fuel consumption. For the same aircraft, utilization rates and block speeds vary by airline and route, so these factors are beyond the direct control of manufacturers and design engineers.

However aircraft productivity and efficiency are measured, they can be improved through advances in aircraft aerodynamics, materials, structures, and other disciplines that improve performance parameters such as lift-to-drag ratio (L/D), ratio of empty weight to MTOW, and specific fuel consumption. Technological approaches to the above goals include the use of boundary layer control to reduce profile drag and parasite drag and the use of new materials, such as modern carbon-based or metal matrix composites, to reduce structural weight fraction. Additional technical areas that merit focused research include composite structures with the following characteristics:

- high damage tolerance
- high stiffness (because a lot of airline structures are sized for stability)
- active controls (which, if sufficiently reliable, may reduce the need for high stiffness)
- low-cost raw materials and fabrication methods
- low density (with high strength-to-weight ratios)
- means to assure that material properties are satisfactory following repairs and have not degraded unexpectedly over the life of aircraft in which composites have been incorporated
- modularity
- resistance to lightning strikes (an area where Boeing, for one, is investing millions of dollars)

Improvements in performance parameters not directly related to aircraft productivity and efficiency are also important, because they would improve the performance of the overall air transportation system. For example, reduced landing and takeoff distances enable aircraft to use more runways (and access more airports). In the extreme, rotorcraft are able to operate without runways. In addition, reducing runway occupancy time during landing and takeoff increases runway throughput. Minimizing the ground footprint of aircraft relative to their capacity also allows for more efficient use of limited airport space.

Although aircraft performance is important, systemwide performance is the overriding concern. Without a systemwide perspective, research and development runs the risk of suboptimization—for example, by improving the performance of a vehicle system in a way that degrades overall performance of the air transportation system. The above discussion of aircraft performance should, therefore, be understood in the larger context of air transportation system performance.

In the discussion that follows, improvements in aircraft performance will be discussed in terms of (1) environmental considerations, (2) advanced airframe concepts, (3) advanced propulsion concepts, and (4) the potential of a cross-cutting technology of particular interest: nanotechnology.

ENVIRONMENTAL CONSIDERATIONS

The air transportation system already expends considerable resources to deal with public concerns and government regulations related to the effects of aviation on local and regional air quality, climate change, and community noise. All of these environmental problems will be aggravated by growth in air traffic. Problems related to emissions are abated by propulsion and airframe concepts and technologies that improve aircraft efficiency. However, the rapid growth of demand for air transportation and the growth in capacity have exceeded the rate of improvement of specific fuel consumption, so that over time aviation consumes larger amounts of fuel. The amount of carbon dioxide (CO_2) released in the atmosphere is roughly proportional to fuel consumption, so more CO_2 is being released. The amount of other emissions, such as oxides of nitrogen (NO_x) and particulates, is also increasing even though engines are becoming more efficient and cleaner, producing fewer emissions per pound of fuel burned. Higher engine combustion temperatures tend to improve the efficiency of the propulsion system, but higher temperatures also increase NO_x emissions. The production of specific emissions can be minimized by changes to the combustion cycle and other aspects of engine design, although changes in engine design to reduce one emission might increase the production of other emissions.

Noise can also be reduced by improvements in the design of the integrated aircraft as well as specific changes to the engine and propulsion system. In some cases, noise reduction technologies reduce overall aircraft efficiency because, for example, they increase aircraft weight.

Breakthroughs could be achieved through use of an alternative fuel such as liquid hydrogen or revolutionary technologies such as fuel cell-electric propulsion. However, breakthrough technologies such as these are likely to take several decades, at least, to become operational. Accordingly, research in environmental technologies should focus on conventional jet propulsion systems, while continuing to explore promising longer-term technologies. Environmental considerations are discussed in more detail below and in a recent report by the National Research Council (NRC, 2002a).

Local and Regional Air Quality

The principal concerns regarding local air quality are high levels of NO_x and particulate matter. At a regional level, NO_x and unburned hydrocarbon emissions from aircraft engines also contribute to the formation of ozone and are currently regulated in accordance with standards established by the International Civil Aviation Organization. A standard for measuring particulate matter from aircraft engines is currently being developed. The contribution of aircraft to regional emissions of NO_x and particulate matter is currently on the order of 1 percent of all anthropogenic emissions. The aircraft contribution is increasing, however, as air traffic increases, while emissions from other sources are decreasing as a result of more stringent emissions standards and improved emissions reduction technologies.

Limits on total NO_x emissions established by local authorities are already threatening to limit capacity at some airports in Europe, while the imposition of landing fees proportional to NO_x emissions by each aircraft have been implemented at other European airports. Stringent emissions standards and the threat of emissions caps have led to modest emissions reductions through optimization of current gas turbine emissions technology. However, these reductions have been largely offset by higher engine pressure ratios (for improved fuel efficiency), which tend to increase emissions of NO_x and particulate matter. Emissions of NO_x by aircraft have not been reduced as much as emissions by surface sources because alternative fuels and exhaust gas cleanup technologies used in other transportation and industrial sources require large, heavy devices that are not practical in aircraft applications. Design improvements that reduce aircraft weight or improve aircraft and engine aerodynamics tend to reduce NO_x emissions because less fuel is consumed.

Better dispersion models will lead to a better understanding of the impact of aircraft emissions and will be a better guide to technology development. Modeling the movement of emissions released by aircraft in flight is significantly more difficult than modeling emissions from a static point source, such as an industrial facility. Another emerging need is the development of a standardized method for measuring emissions of particulate matter; current data on aircraft emissions of particulate matter are sparse and of questionable quality.

Research is needed to develop combustor technologies to reduce emissions of NO_x and particulate matter in engines that operate at high pressure ratios with current jet fuels. If hydrogen fuel becomes widely used in the longer term, particulate matter will no longer be an issue, but low NO_x combustion technology tailored for hydrogen fuel engines will have to be developed.

Climate Change

The aircraft emissions with the strongest effects on climate are CO_2, NO_x, and water vapor (through the formation of contrails and clouds). CO_2 is the most prevalent and best understood greenhouse gas, and the warming effect of CO_2 emitted by aircraft in flight is indistinguishable from that of CO_2 emitted at ground level. NO_x emitted at altitudes normally used by subsonic aircraft forms ozone that can lead to additional warming. The magnitude of this effect is uncertain, but some researchers estimate the warming effect to be two to three times that of CO_2 emissions from aircraft. There is even more uncertainty surrounding the effect of water vapor emitted by current engines; it may be more important—or less important—than CO_2 emissions. Particulate matter emissions can also contribute to warming, though not as much as CO_2 (IPCC, 1999).

Current efforts by the International Civil Aviation Organization to control emissions of CO_2 are focusing on developing a system of market-based options, such as (1) charges based on aircraft efficiency or (2) emissions trading systems. These measures are intended to increase incentives to improve fuel efficiency. However, minimizing fuel consumption might not be the best approach. For subsonic aircraft, global warming effects could be reduced by designing engines to operate at a lower pressure ratio and by designing aircraft to fly at a lower altitude. Both approaches would significantly increase fuel consumption, but based on current understanding of the atmosphere, NO_x emissions and contrail/cloud formation would be reduced enough to offset the increased CO_2 and provide a net reduction in the aircraft's adverse effect on climate. Better understanding of aviation's effect on climate is needed to have confidence that climate change can really be minimized with this kind of strategy (Green, 2002).

Development of combustors that produce ultra-low levels of NO_x at cruise conditions would allow NO_x to be reduced from current levels without increasing fuel consumption (and other types of emissions). On balance, it might turn out that the most beneficial approach would be to optimize the aircraft engine combustors for low NO_x at cruise conditions, even if this increased NO_x emissions in the vicinity of airports. Better understanding of local, regional, and global effects is needed to select the approach that provides the best net environmental benefit.

The use of hydrogen as an aircraft fuel could eliminate emissions of particulate matter and open up new approaches for reducing NO_x emissions, but emissions of water would be increased greatly.

Community Noise

Even if all emissions are reduced to insignificant levels, community noise could limit airport capacity. Many airports already have noise quota systems that limit the number of operations at certain times of each day. The certification process for new aircraft consists of measuring system noise at three points (takeoff, side line, and approach) and is the only universally accepted noise standard for aircraft noise (because it is used for certification). In order to keep objectionable noise within airport boundaries, more advanced engine and aircraft noise reduction technologies are essential. New operational procedures (for example, curved approaches) would also be beneficial at many airports. The goal is to reduce noise below objectionable levels even as traffic increases. Achieving this goal requires an accurate measure of what noise level is objectionable. The FAA currently uses 65 dB DNL[1] as the level that justifies corrective action, but some environmental groups and national governments maintain that the standard should be 55 dB DNL, and complaints about aircraft noise are unlikely to disappear altogether until someone discovers a silent propulsion system. Efforts to reduce objectionable noise are also complicated by the perception (if not the reality) that some noise complaints are motivated less by the level of noise per se than by the fear that aircraft noise causes in those who dread the thought of aircraft routinely flying over their homes. Noise limits may become more stringent in Europe than in the United States. Since U.S. manufacturers compete for sales worldwide, they would need to meet the more stringent noise requirements.

AIRFRAME CONCEPTS

The Vehicle Systems Program of NASA's Aerospace Technology Enterprise has established five vehicle classes to facilitate trade studies of candidate technologies and assessments of technology integration issues.

- personal air vehicles
- uninhabited air vehicles (UAVs)
- supersonic aircraft
- runway-independent air vehicles
- subsonic transports
- other airframe concepts

[1]The FAA uses day-night average sound level (DNL) as a metric, in units of decibels (dB), for assessing annoyance from aircraft noise. It is assumed that operations occurring at night are more annoying than those occurring during the day because they can disturb sleep and because background noise is lower at night. Therefore, DNL is weighted to count each takeoff or landing between 10 p.m. and 7 a.m. the same as 10 daytime takeoffs or landings of equal loudness.

The committee considered how research and technology development applicable to each vehicle class might address the key long-term challenge facing the air transportation system, which is how to accommodate increased demand for air travel while still meeting public expectations related to safety, security, capacity, environmental compatibility, and consumer satisfaction. Technologies related to personal air vehicles, UAVs, and supersonic aircraft were examined by focused NRC studies (2002b, 2000, and 2001, respectively). As described below, personal air vehicles and UAVs are unlikely to contribute significantly to the effort to meet increased demand, although research in both areas would help achieve other goals. For example, improved personal air vehicles could expand opportunities for air transportation to small communities, and UAVs are already fulfilling important military missions.

The ability of supersonic aircraft to help meet increased demand is also problematic, especially in the case of supersonic business jets, which are likely to be the next class of supersonic aircraft to be developed. The development of supersonic business jets may be justifiable in terms of economics and their ability to make service more convenient. However, they are not likely to capture an appreciable fraction of the commercial passenger market, even if technology is available to reduce the sonic boom to acceptable levels for overland travel.

Runway-independent air vehicles may be able to help meet increased demand at capacity-limited airports, and in any case they execute unique, important missions that other types of aircraft cannot perform. However, as in the case of supersonic transports, runway independent air vehicles are unlikely to capture an appreciable fraction of the market for commercial air transportation.

The potential of subsonic transports to meet increased demand far exceeds that of the other vehicle classes. This is not surprising, because the primary purpose of subsonic transports is the efficient mass movement of passengers and cargo. Subsonic transports benefit from design efficiencies unavailable to small aircraft, and they avoid design penalties associated with specialized capabilities such as supersonic cruise speed and vertical flight.

Technologies specifically related to personal air vehicles, UAVs, supersonic aircraft, or runway-independent air vehicles do have the potential to improve the performance of the air transportation system, especially in niche areas. However, research in these areas will not be able to resolve the overall capacity problems that are the primary challenge to the continued success of the air transportation system over the long term. Accordingly, the committee did not examine technologies related to these vehicle classes and makes no recommendations concerning the future direction of research in these areas. Nonetheless, implementation of the process for change recommended by the committee (see Recommendation 5-1) would facilitate planning of research for all vehicle types.

Personal Air Vehicles

NASA's Small Aircraft Transportation System (SATS) project is a key part of NASA's efforts to develop improved personal air vehicles and related airspace technologies and systems. NASA envisioned that the SATS project would relieve some of the capacity problems at the nation's major airports. The NRC determined, however, that it would be very difficult for the proposed SATS concept to address capacity problems. Rather, improved personal air vehicles should be viewed as a complement to commercial carriers that could enhance mobility, especially in regions not well served by scheduled air service. The National Research Council recently completed an in-depth assessment of the SATS project (NRC, 2002b), and another report evaluates research supported by SATS as part of a larger assessment of NASA's aeronautics research (NRC, 2003).

Uninhabited Air Vehicles

The state of the art of UAVs is rapidly advancing, with the Department of Defense investing heavily in UAV research with military applications. UAVs also have potential commercial applications as, for example, "suborbital satellites"—long-endurance aircraft operating at high altitudes (on the order of 100,000 ft) over a fixed location to provide services now provided by satellites. Lighter-than-air vehicles cannot work at these altitudes because available solar energy is insufficient to overcome wind forces. UAVs also have potential as cargo carriers, but this application requires continued research on operational certification requirements for UAVs and changes to air traffic control regulations and procedures. Performance requirements for surveillance aircraft (a key UAV mission for military applications) are very different than for station keepers, so NASA research in this area seems worthwhile. Nonetheless, UAVs, like personal air vehicles, are unlikely to significantly enhance the ability of the U.S. air transportation system to accommodate increased demand.

Supersonic Aircraft

Deployment of an environmentally acceptable, economically viable commercial aircraft capable of sustained supersonic flight, including flight over land, would be a remarkable achievement requiring remarkable technological advances. One approach to the ultimate goal of developing a large commercial supersonic transport would be to first develop a business jet certificated for supersonic flight over land. Development of a supersonic business jet would help address many of the technical challenges involved in developing larger commercial supersonic aircraft and would resolve current uncertainties about the regulatory standards for emissions and noise (including sonic boom) that future commercial supersonic aircraft would be required to meet. Also, some of the technical issues associated with commercial supersonic flight would be easier to address with a smaller aircraft, and economic viability would be easier to achieve for a business jet, which faces different economic drivers than commercial passenger jets. Speed sells, especially in the business jet market, and a commercial supersonic business jet that is able, in terms of technical performance *as well as* regulatory authorization, to cruise at Mach 1.6 to 1.8 over land would probably be a commercial success even at twice the price of a comparably sized conventional business jet (NRC, 2001).

Even though the effort required to overcome the technological barriers to a supersonic business jet is less than that for a large commercial transport, it would still be sizeable. The most significant challenge involves defining acceptable levels of sonic boom and demonstrating that those levels can be met, to support regulatory changes to permit supersonic flight over land. This will probably require building a full-scale demonstrator, which could cost $1 billion.

The NRC issued reports assessing NASA's commercial supersonic research in 1997 and 2001 (NRC, 1997, 2001). The more recent report (1) identifies key technology challenges for supersonic business jets and two classes of larger commercial supersonic transports and (2) makes specific recommendations for research. A properly directed and adequately funded research and technology effort could probably enable operational deployment of environmentally acceptable, economically viable commercial supersonic aircraft in 25 years or less—perhaps a lot less if there is an aggressive technology development program for aircraft with cruise speeds less than approximately Mach 2 (NRC, 2001).

For a given payload, range, and MTOW, a high-speed subsonic aircraft will have higher productivity and efficiency than a slower aircraft. However, the ability to cruise at supersonic speeds is not without cost: Supersonic flight increases specific fuel consumption and requires a more robust airframe design. As a result, a supersonic aircraft has a higher fuel weight fraction and a shorter range and/or higher MTOW than a subsonic aircraft with a comparable payload capacity. It is far from certain whether a commercial supersonic aircraft would be more efficient or have higher productivity than subsonic aircraft, and the committee is not aware of any research that characterizes commercial supersonic aircraft as a solution to increased demand for air transportation. More commonly, support for commercial supersonic aircraft is based on other important factors: their ability to provide better service (by reducing travel time), the national economic benefits from being first to market with a commercially successful supersonic aircraft, and the economic damage from a foreign aerospace company being first to market.

Runway-Independent Air Vehicles

Rotorcraft are an essential part of the air transportation system. They provide access to disaster scenes, oil rigs, hospitals, maritime vessels, building rooftops, construction sites, and other locations that other forms of aviation cannot service. With regard to the broad challenge of meeting the general public's increased demand for air transportation, one of the great potential payoffs offered by rotorcraft—or other commercial aircraft with vertical takeoff or landing (VTOL) capabilities—is at airports that are operating near or at their capacity limits. VTOL aircraft have the ability to provide passenger service without increasing demand for runway usage. Currently, the operating cost per seat-mile for VTOL aircraft is 4 to 10 times higher than for conventional aircraft. In the near term, the commercial success of a greatly expanded network of VTOL aircraft would require economic incentives or subsidies to offset their higher cost. In the long term, technology could help reduce the cost differential and address other issues, such as operation in adverse weather (especially icing conditions) and the noise of rotorcraft operations.

One way for NASA to stay involved in research related to rotorcraft and other runway-independent air vehicles would be through partnerships with the Department of Defense, which has invested heavily in this area.

Subsonic Transports

Nontraditional concepts for new classes of commercial aircraft have the potential to greatly improve the performance of both small and large subsonic transports. Concepts of particular interest include (1) the strut-braced or joined wing and (2) the blended-wing-body (BWB) (see Figure 4-1).

Strut-Braced or Joined Wing

The joined wing configuration is similar in concept to a biplane whose lower wing has positive dihedral (the tip of the wing is higher than the point of attachment to the aircraft centerbody) and is swept back from where it is attached to the fuselage. The upper wing is attached to the vertical tail, sweeps forward with negative dihedral, and is joined to the lower wing. A plan view shows the wing as a rhombus, which gives the wing a rigid structure and allows it to have a greater span.

A strut-braced wing uses a strut to support the wing, allowing increased aspect ratio, reduced wing thickness, lower weight, reduced wing sweep, larger wing areas with laminar flow, reduced drag, higher L/D, smaller engines, reduced fuel consumption, and reduced noise and emissions. Pfenninger (1987) has estimated that a strut-braced wing configuration could provide an L/D of 40 if laminar flow

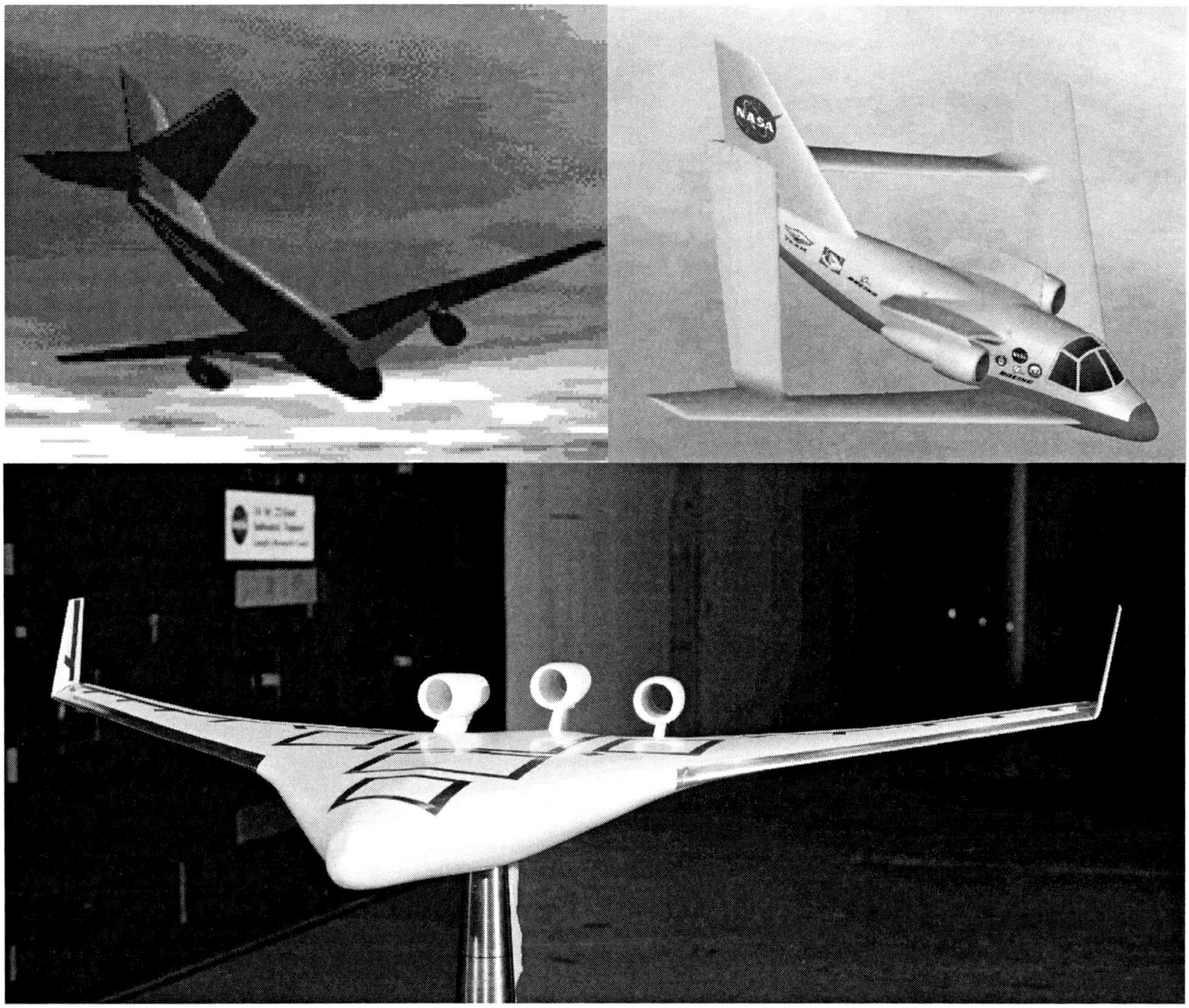

FIGURE 4-1 Nontraditional aircraft concepts: strut-braced wing (top left), joined wing (top right), and blended-wing-body (bottom). Source: NASA.

boundary control is applied. A NASA-sponsored study compared the performance of various strut-braced wing designs with the performance of a traditional design with a cantilevered wing. The study concluded that, for an aircraft with a capacity of 325 passengers and a service entry date of 2010, strut-braced wing designs would have a lower takeoff gross weight (by 9 to 17 percent) and lower fuel consumption (by 14 to 22 percent) (Gundlach et al., 1999).

Blended-Wing-Body Aircraft

BWB aircraft are a form of flying wings, a configuration that has been investigated by many different aircraft designers, manufacturers, and countries for over 50 years. During the 1940s, the United States developed flying-wing bomber prototypes (the propeller-powered YB-35 and a jet-powered variant, the YB-49); the Germans supported development of flying-wing bombers and fighter-bombers; and the United Kingdom developed a flying-wing fighter prototype, the Armstrong Whitworth AW-52. The Hawker Vulcan is a successful flying-wing-type aircraft that had the same payload range characteristics as the B-47. Currently, the most notable flying-wing aircraft is the U.S. Air Force B-2 bomber, which first flew in 1989.

BWB aircraft integrate the fuselage into the wing structure to reduce fuselage drag. This concept was suggested for commercial applications by Boeing about 10 years ago and has been the subject of studies by Boeing, NASA, and universities in the United States and Europe. The thick center section distinguishes the concept rather fundamentally from pure flying wing designs and leads to structural efficiencies that improve the overall performance of BWB aircraft.

A BWB aircraft with an optimized design could have an L/D as much as 50 percent greater than that of current aircraft. Performance could be further improved by the ability to use turboprop propulsion and the option of hydrogen fuel (because of the large size of super cargo aircraft, the low density of hydrogen fuel becomes less of an issue than it is with conventional aircraft). Existing commercial cargo aircraft are derivatives of passenger aircraft or military cargo aircraft. Technological advances would allow a modern cargo aircraft to have much better performance than multiple-use aircraft. Laminar flow over the wings would further increase the efficiency of BWB aircraft—and of every other aircraft type discussed above.

Two major airlines have indicated interest in purchasing BWB aircraft if they are ever developed. However, such expressions of interest—no matter how vigorous—do not always result in a commercially successful product. For example, the GE 36 was an unducted fan engine that completed flight demonstrations in the late 1980s (see Figure 4-2). The engine consumed 35 percent less fuel than conventional engines with comparable performance, and during initial development airline representatives responded very favorably. When it came time to make a binding commitment to purchase the engines, however, airlines decided that greatly improved fuel economy was insufficient to overcome concerns about life-cycle costs, noise, blade loss, and the possibility that airline passengers might be put off by the appearance of the engine's external blades. Concerns that BWB aircraft will have to overcome include passenger acceptance of cabins with few, if any, windows; how to handle emergency evacuation of large passenger cabins; and how to fit such a large aircraft (in terms of passenger capacity and size) into existing airport environments.

FIGURE 4-2 Unducted fan demonstrator ready for flight. Source: Burkhard Domke, Grünendeich, Germany. Available online at <www.b-domke.de/AviationImages/Rarebird/0809.html>.

Other Airframe Concepts

The committee considered other vehicle concepts, as well, but does not recommend focused technology research related to these concepts.

Wing-in-Ground-Effect Aircraft

Wing-in-ground-effect aircraft were pioneered in the former Soviet Union. In 1967, the Defense Intelligence Agency detected a Soviet wing-in-ground-effect aircraft with a MTOW of more than 1 million pounds operating in the Caspian Sea (Losi, 1995).

Ground-effect aircraft fly at an altitude equal to a fraction of the wingspan in an aerodynamic regime called "ground effect," which increases aircraft efficiency. According to some mathematical models, the aerodynamic efficiency of a wing-in-ground-effect aircraft continually increases as the altitude of flight decreases. Operational and safety considerations, however, will always define a minimum flight altitude. Because of their low flight altitude, wing-in-ground-effect aircraft are generally unsuitable for operation over land, though they may be feasible in arctic or desert areas. To operate over open ocean, wing-in-ground-effect aircraft have very large wingspans to avoid collisions with waves. Even with a MTOW on the order of several million pounds, the cruising altitude would be so low that a wing-in-ground-effect aircraft might be at risk from rogue waves.[2] The vast size of such an aircraft means it would take a tremendous financial investment to produce an operational product. Also, the power required during takeoff is much greater than the power needed for cruise, thereby increasing the mass of the engines and reducing both payload and aircraft efficiency (Losi, 1995).

Lighter-Than-Air Craft

Extremely large lighter-than-air craft are another possibility for improving the productivity and efficiency of commercial aviation. Diesel propulsion becomes a possibility for lighter-than-air craft, but water vapor in the exhaust would need to be captured to control the buoyancy of long-distance transports.

Seaplanes

The feasibility of seaplanes is limited; they require more propulsion power than a similarly sized conventional aircraft to take off, and there are relatively few landing sites. Protected harbors with ready access to cargo facilities that could accommodate a seaplane facility generally are already

[2]Rogue waves result from a superposition of waves in the open ocean, which causes a large wave to rise up from the surface of the ocean.

tied up for other uses. The integrity of takeoff and landing areas is also an issue; floating logs and other debris are a major hazard.

Airframe Research Needs

Additional airframe research and technology development are needed to improve the performance of aircraft, particularly with regard to the feasibility of nontraditional subsonic transport concepts. For example, improved materials, especially composite materials, have the potential for tremendous payoffs. The economic viability of many nontraditional concepts, especially the BWB, would be enhanced by composite materials that weigh less, are more damage tolerant, are easier to fabricate, are more suitable for modular construction techniques, and are compatible with effective means for joining assemblies.

Other areas of particular interest include technology transition issues, safety, and security. To better understand the research requirements related to each of the vehicle classes associated with NASA's Vehicle System Program, NASA could form a team for each class with members from government, industry, and academia. If such teams are established, a method should be established to ensure that cross-functional performance and design issues, such as reduced wake vortices, reduced drag, improved high-lift performance, and reconfigurable wings, also receive adequate attention.

In some cases, NASA is constrained by the administration or the Congress from supporting aeronautics research related to an aircraft concept of particular interest to an aircraft manufacturer; the concern is that the government may inappropriately subsidize industrial research. The danger is that NASA research may be limited to topics of little or no interest to industry, which brings into question the value of conducting the research.

PROPULSION CONCEPTS

The committee reviewed a range of potential aircraft propulsion cycles and configurations in an effort to assess how the NASA vision and goals might be addressed in the specified time periods. In this process the committee recognized the importance of evaluating the entire propulsion package in terms of the weight, volume, and costs of the prime movers and the related fuel. The committee also recognized the importance of evaluating these parameters in the context of the overall performance of aircraft and the transportation system—not just as a vehicle subsystem. Figure 4-3 compares the conventional heat engine cycles assuming component efficiencies of 100 percent and maximum hydrocarbon fuel combustor temperatures. As shown, the simple Brayton

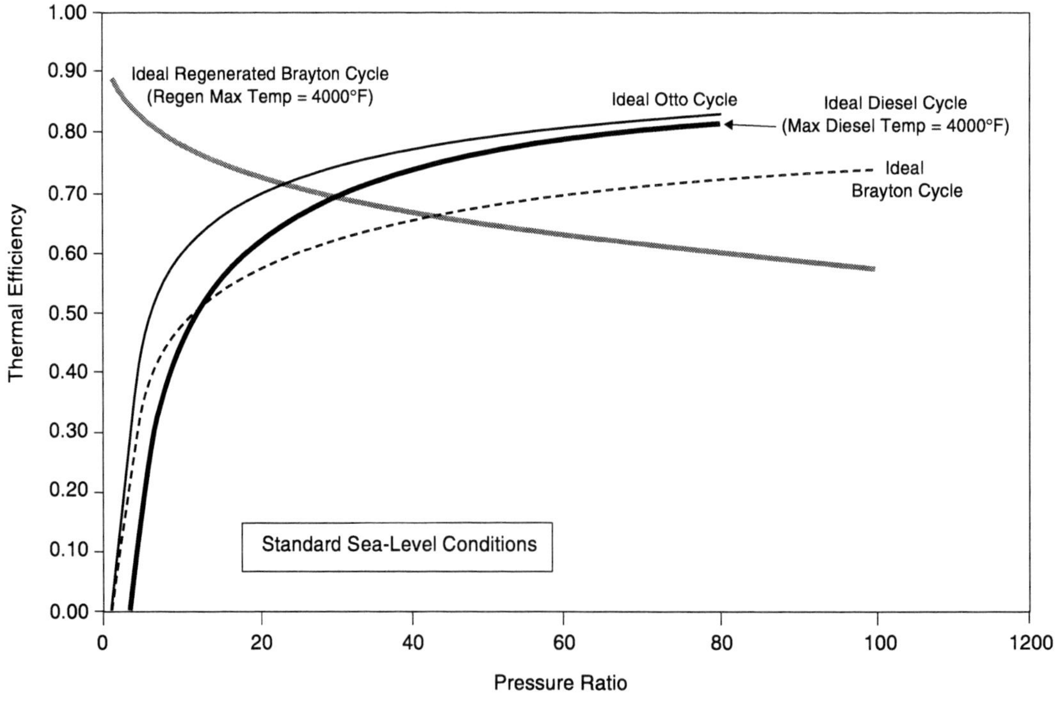

FIGURE 4-3 Thermal efficiency versus pressure ratio for conventional heat engine cycles assuming component efficiencies of 100 percent and maximum hydrocarbon fuel combustor temperatures. Source: Jeffrey M. Stricker, Wright-Patterson Air Force Base, Aero Propulsion Laboratory, briefing to committee members S. Michael Hudson and Willard J. Dodds, January 13, 2003.

cycle, while not providing the absolute maximum cycle efficiency is very competitive on this "ideal" basis. Furthermore, the Brayton cycle operates continuously, whereas the Diesel and Otto cycles operate intermittently, so their net performance is not nearly as good as their ideal performance. As a result, the Brayton cycle proves to be superior to the Diesel and Otto cycles for many applications. In aviation, the Brayton cycle has historically been the cycle of choice when the effects of propulsion system weight, volume, and durability are factored into the entire aircraft and air transportation system.

The efficiency of the Brayton cycle is governed by the maximum compressor exit temperature, which is determined by high temperature material limits. The propulsion taxonomy in Appendix D describes a broad range of propulsion concepts and substantiates the conclusion that the conventional gas turbine engine and its variants based on the Brayton cycle will continue to be the primary aircraft propulsion system of choice at least through 2025. One of the variants entails a departure from the concept of isentropic compression and expansion. Here, two alternatives exist. The first is combustion in the turbine (i.e., interturbine burning). This alternative offers the possibility of reducing the temperature drop across the turbine while increasing the work done by the turbine, thus improving overall engine performance even if interturbine burning is active only during peak power (takeoff and climb). A second alternative is the introduction of volume cooling ahead of or in the compressor by introducing a mist of water or other coolant either ahead of or between compressor stages. This alternative has two benefits. The first is the increase in total pressure owing to the volume cooling, and the second is the increase in mass flow. The former has the effect of increasing the compressor efficiency in that the compressor outlet temperature (T_3) is reduced for a given pressure ratio. The latter increases the exit momentum flux, which could also be used to increase takeoff and climb performance. Either improvement could reduce the propulsion system weight fraction and improve aircraft efficiency. These modified Brayton cycles warrant research and could be incorporated into operational systems by 2025.

Turbomachinery-Based Propulsion Systems

Propulsion system performance is directly related to the safety, capacity, mobility, noise, and emissions of individual aircraft and the air transportation system as a whole. Figure 4-4 shows the significant advances that have been made in

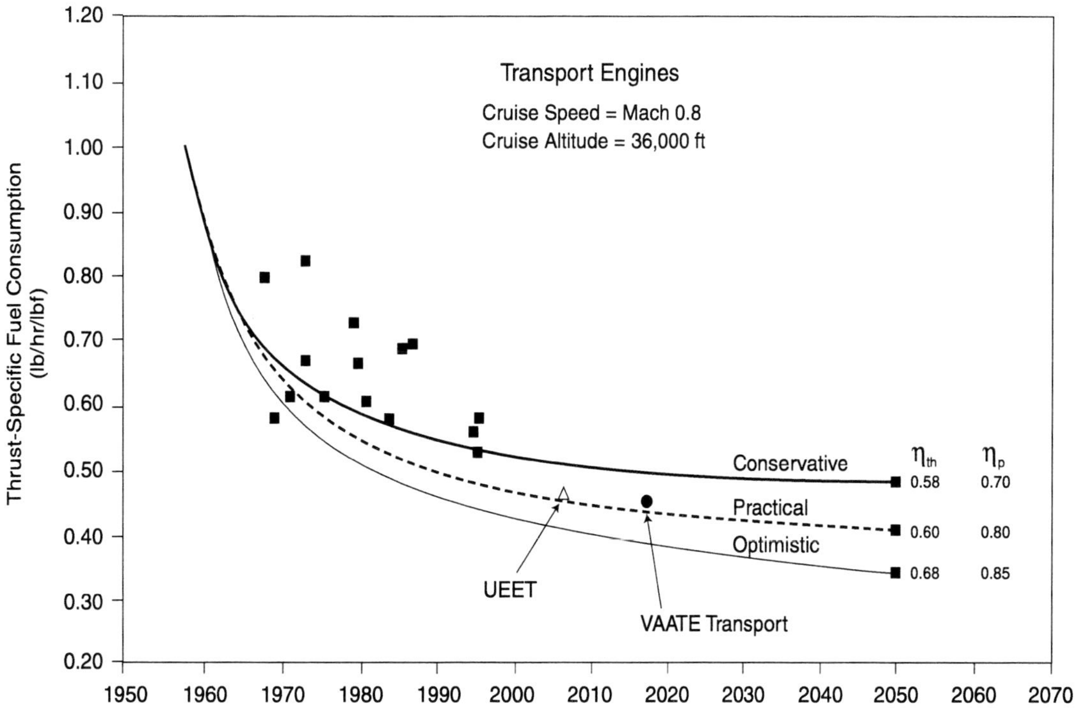

FIGURE 4-4 Predictions made in 1968 of subsonic thrust-specific fuel consumption, updated with data on operational systems developed since 1968. UEET, ultra-efficient engine technology; VAATE, versatile, affordable, advanced turbo engines; η_{th}, thermal efficiency; η_p, propulsive efficiency. Source: Jeffrey M. Stricker, Wright-Patterson Air Force Base, Aero Propulsion Laboratory, briefing to committee members S. Michael Hudson and Willard J. Dodds, January 13, 2003. Modification of data from L. Dawson. Propulsion. Aeronautical Journal of the Royal Aeronautical Society 72(September):209-229.

gas turbine propulsion systems using thrust-specific fuel consumption as a figure of merit. Although the rate of improvement in this commonly used parameter has decreased, incentives remain great to pursue further technology advances in order to meet customer-driven goals. As an example, NASA has successfully worked with industry to develop and verify analytical tools that address design and system evaluation and reduce the number of experiments needed for a successful engine.

Commercial aviation is highly competitive, so minimizing costs is critical to the survival of individual airlines. The propulsion system is not immune to these pressures: Advanced propulsion technologies will rarely be incorporated in operational products unless they reduce costs or are needed to meet some other requirement, such as more stringent noise or emissions standards.

The propulsion systems of commercial aircraft are only a small contributor to the accident rate as a result of tremendous investments and decades of work to improve the reliability of turbomachinery. It is essential that new propulsion systems and components also demonstrate very high levels of safety.

Propulsion research plans should be structured to meet the needs of advanced airframe concepts in the context of the long-term vision for the air transportation system. Concepts such as BWB aircraft, supersonic business jets, and runway-independent aircraft dictate unique requirements and opportunities for advances in propulsion technology. Areas of interaction include extremes in engine size and fan bypass ratio, design for boundary layer ingestion, highly integrated engine-airframes, power extraction for boundary layer manipulation, variable cycle features, and architectures for integration of system controls.

Emerging Propulsion Concepts and Fuels

In the 2025 to 2050 time frame, low-cost hydrogen could become attractive as an aircraft fuel that would reduce the environmental effects of aviation. The key challenge to the use of hydrogen as an aircraft fuel is its low energy density compared with hydrocarbon fuels—unless new (high-density) means of storing hydrogen are developed. Even though the committee is not aware of any particularly promising approaches for overcoming this problem, the high potential payoff warrants continued research. Widespread use of hydrogen as an aircraft fuel would also require an economically and environmentally benign method for producing hydrogen, a challenge that is being addressed by broader efforts to enable hydrogen to replace hydrocarbon fuels in ground-based vehicles and industrial uses.

Advances in electric power systems may eventually allow them to replace internal combustion engines. In particular, methane or hydrogen fuels for fuel cell power systems in various forms offer a potentially significant improvement in energy conversion efficiency over today's gas turbines, and ongoing research programs are addressing both mobile and stationary fuel cell applications. Even so, tremendous advances in the power density of fuel cells would be required to make them technologically feasible as a source of propulsion power for large commercial aircraft. Other technology issues associated with the development of an electric aircraft propulsion system (such as the development of lightweight electric motors using, for example, room temperature superconductors) would also need to be resolved to make a fuel cell energy conversion system into a successful aircraft propulsion system. It might also be feasible to use the electricity produced by fuel cells to add heat to the gas in a gas turbine engine in place of combustion. Since electricity, lasers, and electromagnetic devices can provide volumetric heating in place of combustion of hydrogen or hydrocarbon fuel, exploratory research is in order to determine the conditions under which these alternatives may be attractive. Other alternative approaches are given in Appendix D.

The first application of electric power sources on commercial aircraft is likely to be as auxiliary power units rather than for propulsion. Although fuel cells are larger and heavier than conventional auxiliary power units, they generate water. This would reduce the amount of water that must be carried onboard at takeoff, thereby improving the overall assessment of fuel cells as auxiliary power units, from a systems perspective. Unrelated technology developments, however, may produce aircraft toilets that flush with 90 percent less fluid, reducing the onboard demand for water.

Intermittent combustion concepts, such as pulse jets (a.k.a. pulse detonation engines), have the potential for improved performance relative to traditional turbomachinery systems. In some cases, intermittent concepts may also significantly reduce complexity. However, it seems unlikely that systems based on intermittent concepts will outperform gas turbine aircraft propulsion systems in the foreseeable future. Nonetheless, continued basic research would be worthwhile to better understand the long-term limitations and potential benefits of intermittent combustion concepts.

Nuclear power is unsuitable for aircraft applications for many reasons, including the weight of radiation shielding, radiation exposure during normal operations, and the risk of widespread radioactive contamination in the event of an accident. The committee did not identify any other specific propulsion or fuel concepts of particular interest, although research to explore new concepts would be consistent with NASA's vision and goals.

Propulsion Research Needs

Future airframe and propulsion research will lead to a better understanding of the synergies and tradeoffs that exist among system and subsystem concepts, technologies, design characteristics, and performance parameters, including environmental performance parameters—for example, specific fuel consumption, noise, and specific engine emissions. Currently available information indicates that propulsion research should generally support the continued evolution and use of high-bypass turbofan engines burning liquid hydrocarbon fuels. At the same time, a portion of the research should anticipate the possibility of (1) an eventual changeover to hydrogen, (2) the use of an advanced gas turbine engine core, and/or (3) the use of fuel cells to generate electric power for electrically driven engines if and when room temperature superconductivity becomes practical. Research in these areas should start at a low level and proceed at a pace consistent with research focused on nonaerospace applications.

The development of environmentally beneficial propulsion technologies that might eventually be applied to aircraft systems should be tracked to understand their potential environmental benefits and tradeoffs (for example, evaluating the potential advantages and disadvantages of using hydrogen fuel, including the potential to use cryogenic hydrogen as a heat sink for electrical components). Support should be provided for research necessary to take advantage of new technologies, including the design of components (such as low-emission combustors compatible with hydrogen fuel or electrically driven propulsion engines compatible with advanced fuel cells) and the development of new system concepts (such as an environmentally acceptable means of releasing water into the atmosphere to mitigate the effect of greatly increased emission of water vapor that would result from the use of hydrogen fuel).

AVIONICS

Avionic systems include computers, communications networks, sensors, controls, operational software, and human-computer interfaces. Avionics plays an increasingly critical role in the safe and efficient operation of commercial aircraft and now accounts for up to 40 percent of the capital cost of new aircraft.

Onboard electronics perform or monitor virtually all critical functions in an aircraft, including engines, control surfaces, stability augmentation systems, active flow controls, flight path, collision avoidance, and interactions with the external air traffic control system.

The federal government, especially the Department of Defense, has supported basic long-term research and applied research and technology development that continue to enhance the capabilities of avionics on both civil and military aircraft. The success of this research has been enabled in large part by smaller, more capable computers and more sophisticated software.

Chapter 2 discussed the importance of research in automation and the ability of automated systems to enhance the performance of human operators and the overall system. Advanced on-board avionics will be necessary to implement any new operational concepts that call for increased automation of cockpit and navigation functions.

Federal agencies should continue research aimed at enhancing airborne avionic systems through evolutionary improvements, while pursuing longer range research that could lead to major breakthroughs. For example, advances in nanotechnology may provide major benefits to avionics in computing, sensors, and active distributed control. Research in avionics that relates to air traffic control and automation should be integrated into the overall research strategy for the air transportation system as a whole. Two examples of ongoing research and development of this type are (1) the Alaska Airlines all-weather approach system and (2) work by NASA's SATS project to enable safe low-visibility operations at minimally equipped landing facilities through the development of new operational concepts, sensors, pilot interfaces, and procedures.

NANOTECHNOLOGY

Nanotechnology is an emerging technology with the potential for broad application to many aspects of aircraft design. Nanotechnology deals with materials and devices having at least one dimension on the order of 1 to 100 nanometers. The design of nanoscale materials deals with molecular-scale structures, whose physical and chemical properties are different from materials at larger scales. At nanoscales, no atom is far from a surface. This changes chemical reactivity, coherent scattering, and other processes. Devices also involve large, but countable, numbers of atoms.

Nanotechnology is still in its infancy and is just starting to move into operational applications. In fact, the term nanotechnology is somewhat misleading, implying that research has generally advanced to the stage of developing useful technology, when in many (or most) cases, nanoscale research is still scientific research (and would more accurately be referred to by the less common term nanoscience).

Global investment in nanotechnologies is about $1.5 billion per year, primarily in the United States, Europe, and Asia. The U.S. federal government appropriated $604 million for nanotechnology research and development in fiscal year 2002. The four agencies most heavily involved in nanotechnology research and development are the National Science Foundation ($199 million), the Department of Defense ($180 million), the Department of Energy ($91 million), and NASA ($46 million) (Roco, 2002).

Areas of Interest

The aeronautical community should maintain an awareness of nanotechnology research in other disciplines that might be used in aeronautical applications, such as flow control, lubrication, structures, and manufacturing. A recent study (NRC, 2002c) on the future role of micro- and nanotechnologies in improving Air Force capabilities identified three scientific frontiers that nanotechnology research should explore: materials, devices, computer processing requirements, and fabrication.

Materials

Research into nanotechnology devices for aeronautics applications should investigate bonding of dissimilar materials, material properties, and scaling. Industry would greatly benefit from any technology that improved the ability to bond dissimilar materials. Microelectromechanical systems (MEMS) research is investigating the ability to create strong bonds between (1) silicon and silicon and (2) silicon and other materials, and the committee is hopeful that nanotechnology research might someday lead to material bonding methodologies for critical aviation applications.

Nanotechnology may lead to the development of new structural materials with high strength-to-weight ratios and fracture toughness, durable coatings, greater resistance to corrosion, self-healing, and multifunctional characteristics. For example, structural materials might have embedded sensors and actuators; custom-designed properties, such as electrical conductivity, mechanical strength, magnetic behavior, and optical properties; or improved damping properties. Multimode damping could lead to the elimination of swash plates in helicopter rotors, which would be a major design breakthrough, and greatly reduced fatigue failure in turbofan blade applications. Self-healing materials (e.g., materials embedded with small particles of liquid that would be released and fill in cracks to prevent them from propagating) may allow flying aircraft closer to their fatigue limits, but generally the benefits of self-healing are likely to be greatly exceeded by the benefits of increased strength and reduced weight.

The properties that nanomaterials demonstrate at nanoscales do not necessarily predict the properties of macroscale materials that incorporate nanomaterials. For example, nano-microtubes have heat-transfer rates comparable to that of diamonds, but more research is needed to assess the ability of nanotubes to increase the heat transfer capabilities of structural materials. Also, segments of some nanotechnology fibers are on the order of 30 times stronger than glass fibers. The challenge is to demonstrate strength on a macroscale by combining strong nanoscale segments to form suitable matrix composite materials.

Devices

Research into nanotechnology devices for aeronautics applications should investigate distributed sensing, electric propulsion, flow control, fuel controls, MEMS materials, photonics, and security.

Nanotechnology can support distributed sensing: adhesive tape with embedded sensors has been developed that can be used on vehicles during flight tests. In the future, distributed sensors may transition from research applications to operational applications, where they would be used as part of the flight control system.

The feasibility of electric propulsion would be enhanced by the development of (1) a fuel cell catalyst that would not spoil (or become poisoned) as current catalysts do during the operating process or (2) a room-temperature superconductor.

An application area with near-term potential is flow control. Long term, nanotechnology has a role to play in reconfigurable wings. (A NASA flight test recently demonstrated wing warping.) Potential benefits might include the elimination of moving control surfaces, resulting in hingeless wings, the ability to adjust wing camber in flight to reduce drag and improve lift, improved handling qualities and maneuverability, and reduction of noise and vibration.

Propulsion efficiency could be enhanced through the development of better fuel controls (e.g., sensing and calculating devices to measure fuel flow, temperature, and pressure).

Nanoscience is the key to developing materials for MEMS devices, particularly with regard to structural stability, surface durability, fabrication, and packaging (NRC, 2002d). For example, an aircraft skin embedded with MEMS devices might greatly reduce turbulent skin friction.

Optics is used for transmission of data over long distances. The Defense Advanced Research Projects Agency (DARPA), among others, is conducting research to use light in processing on an integrated circuit. Light offers great advantages—and creates large challenges. The photons being used are larger-than-nanoscale, on the order of a half-micron. This is much larger than the transistors they would replace, so photon-based integrated circuits would be larger than current devices.

Carbon nanotubes arranged in sensor configurations might contribute to the development of more capable explosives detectors. (Carbon nanotubes are small graphite cylinders with unusual electrical properties. They can act as metals, semiconductors, or insulators, depending on how they are constructed.)

Computer Processing Requirements

Deployment of distributed nanotechnology sensors will require significant advances in computer processing, so that data from hundreds or thousands of sensors can be processed in real time. Similar challenges would be associated with the use of swarms of small autonomous vehicles based on micro- and nanotechnology. Investments in research and development for algorithms, architectures, and software are necessary to maximize the utility of new hardware (NRC, 2002d).

Fabrication

Developing nanomaterials and devices will require research into the assembly of multifunctional nanostructures.

A recent study recommended that the Air Force monitor and selectively invest in self- and directed fabrication and assembly, particularly with regard to processes related to primarily military components, such as sensors and propulsion (NRC, 2002c). Current abilities in the self-assembly of nanoscale materials are rather crude, suitable for growing crystals, for example. However, biological organisms are all self-assembled systems, so the potential obviously exists.

Bottom Line

The potential posed by nanoscience and technology is enormous, but how, when, and the extent to which this potential will be realized are impossible to predict, and the specifics of predictions become more uncertain the farther they are extended into the future (NRC, 2002d). It is especially difficult to determine how the application of nanotechnology may improve top-level characteristics such as overall aircraft performance or the safety of the air transportation system. To date, nanotechnology has been very successful in some devices, but not in devices large enough and robust enough to be directly applicable to commercial aviation. Major advances in the application of nanotechnology are likely to depend upon the ability to integrate nanotechnology fibers and features in intelligent ways to create macroscale materials with specific desired properties. The National Science Foundation recently completed a solicitation for research in this area.

To be successful, nanotechnology research and technology must be sustained over the long term. It will take time and money for research in particular areas to bear fruit. Some practical results in the near term would be helpful in sustaining long-term support for nanotechnology research. The National Nanotechnology Initiative is structured to support projects with near-term, mid-term, and long-term applications. With nanotechnology, as with any new technology, industry funding of research and technology will occur only when supported by a solid business case: "Will it make money—or increase market share?" Ultimately, the success of nanotechnology rests upon the development of successful commercial products.

NASA and the aeronautics community should continue their involvement in interdisciplinary nanotechnology research and development to ensure that advances will be applied to applications of interest to aviation. For example, in June 2002 NASA established seven University Research, Engineering and Technology Institutes, including one for bionanotechnology materials and structures for aerospace vehicles at Princeton University and Texas A&M University. Research by the institutes is intended to increase fundamental understanding and lead to the development of basic technology. The institutes will also support the education of university students and training for working engineers and scientists.

Most nanotechnology research and development in the United States is structured using a bottom-up investment strategy, where individual agencies do research in areas of interest to them. To better integrate research plans at a high level, the NRC has already recommended that the Office of Science and Technology Policy establish an independent standing committee to advise the federal interagency coordinating committee for nanotechnology on research investment policy, strategy, program goals, and management processes. Nanotechnology research and development in the United States would also benefit from the formation of a "crisp, compelling, overarching strategic plan" to articulate short-, medium-, and long-term goals and objectives (NRC, 2002d). Nanotechnology research related to commercial aviation would likely benefit from the implementation of these recommendations.

RECOMMENDATIONS TO IMPROVE AIRCRAFT PERFORMANCE

The Integrated High Performance Turbine Engine Technology (IHPTET) program exemplifies one approach for conducting advanced research on application-ready technology. The IHPTET program was a joint Department of Defense-NASA-industry program whose cost was shared 25 percent-25 percent-50 percent, respectively (with the industry money coming from internal research and development funds, some of which is earned on other government contracts). IHPTET produced useful research results that were transitioned into operational products because systems demonstration at technology readiness level (TRL) 6 was included in the program.[3] Each phase of the IHPTET program had high-level goals (e.g., thrust-to-weight ratio) as well as component-level goals (e.g., efficiency and cooling flows). The process was successful because it allowed different groups participating in the program to develop different approaches to high-level goals, and the lower-level goals depended on the accepted systems approach. As a consequence, IHPTET, as a program, did not try to pick winners in a technical sense. The program's flexibility was a key to its success.

Similarly, a research program to improve the performance of commercial aircraft could be structured with several parallel tracks to cover short-term, medium-term, and long-term goals. Each track would focus on concepts with the potential

[3]NASA uses TRL to define levels of technological maturity. The lower the TRL, the more research and development is needed to prepare a technology for commercial application. TRL 1 implies that basic principles have been observed and reported. TRL 6 implies that a system or subsystem model or prototype has been demonstrated in a relevant environment. TRL 6 is traditionally the level at which NASA considers technology ready for transfer to industry in preparation for commercial product development. TRL 8 means that a system has been flight qualified and is ready to begin operational use.

to reach maturity by a given date. For example, one track might focus on BWB designs with turbine engines and conventional fuels, with the idea of reaching maturity in 15 years; another track might focus on BWB aircraft with laminar flow and conventional fuels, with a maturity date of 2030; and a third track might focus on hydrogen fuel and advanced propulsion concepts, with a maturity date of 2050. This kind of framework would facilitate coordinated research along a wide range of interesting technologies while producing a steady stream of operationally useful technologies.

Finding 4-1. Advanced Aircraft Technology. Improvements in aircraft performance are critical to achieving necessary improvements in almost every aspect of the overall performance of the air transportation system. Innovative long-range research leading to the implementation of new operational concepts is also required for the air transportation system to take full advantage of gains in the performance of commercial aircraft.

Recommendation 4-1. Aircraft Research and Technology. To improve the performance of aircraft through 2025, federal agencies should continue to support research leading to evolutionary improvements in aircraft performance. Looking out to 2050, however, research should support innovative concepts aimed at major advances in performance. In addition, agencies should continue to monitor research in related emerging technologies, such as nanotechnologies, and support research aimed at aircraft applications as emerging technologies mature. The areas listed below are prime candidates for this kind of long-term research.

- Analytical tools, advanced technologies, and the fundamental science behind both, to reduce the need for costly hardware testing and to more easily achieve overall research goals (especially in emerging technologies).
- Composite materials with improved qualities that would increase their use in airplane structures and reduce aircraft weight.
- Environmental consequences of aircraft noise and emissions locally, regionally, and globally, to better understand those consequences and support the establishment of better informed priorities and goals for noise and emissions reduction that (1) reflect the need for integrated approaches (involving advances in airframes, engines, and operational procedures) to meet environmental goals and (2) accurately account for the tradeoffs among different environmental goals and different approaches to achieving those goals.
- Low emissions combustor technology, to (1) reduce substantially emissions of NO_x and particulate matter at airports (to improve local and regional air quality) and at cruise altitudes (to reduce global climate effects) and (2) reduce emissions produced by engines with high pressure ratios.
- Nanotechnology, to explore its long-range potential for dramatically enhancing aircraft performance through the development of advanced avionics (computing, sensors, and active distributed controls) and high-performance materials.
- Nontraditional aircraft configurations, including but not limited to (1) the blended-wing-body and (2) the strut-braced or joined wing, to improve aircraft productivity and efficiency and reduce noise and emissions.
- Passive and active control of laminar and turbulent flow on aircraft wings (laminar flow to increase cruise efficiency and turbulent flow to increase lift during takeoff).
- Nontraditional power and propulsion concepts and technologies, especially concepts and technologies that support the use of alternative fuels, such as fuel cells (which may have application as auxiliary power units in the foreseeable future) and high-density storage of hydrogen to improve the feasibility of using it as a propulsion fuel.
- High-temperature engine materials and advanced turbomachinery, including (1) lower speed, highly loaded, fan drive turbines and fan reduction gears; (2) very large fan systems, which require advances in manufacturing and material systems; (3) boundary layer control on turbomachinery airfoils, to improve component efficiency and packaging; (4) aspirated turbomachinery components, which could greatly reduce noise and improve component efficiency; and (5) other innovative concepts, such as interturbine burning or volume cooling ahead of or in the compressor by means of a mist of water or other coolant.

REFERENCES

Green, J. 2002. Greener by design: The technology challenge. Aeronautical Journal 106(1056):57–113.

Gundlach, J., A. Naghshineh-Pour, F. Gern, P. Tetrault, A. Ko, J. Schetz, W. Mason, R. Kapania, B. Grossman, and R. Haftka. 1999. Multidisciplinary Design Optimization and Industry Review of a 2010 Strut-Braced Wing Transonic Transport. MAD 99-06-03. Blacksburg, Va: Virginia Polytechnic Institute and State University. Available online at <http://www.aoe.vt.edu/research/groups/tbw/reports/jay/mad0603.pdf>.

Intergovernmental Panel on Climate Change (IPCC). 1999. Aviation and the Global Atmosphere. Cambridge, U.K.: Cambridge University Press.

Losi, P. 1995. The Wingship's Potential for Strategic Lift. Executive Research Project. The Industrial College of the Armed Forces. Fort McNair, Washington, D.C.: National Defense University. Available online at <www.ndu.edu/library/ic6/95-S12.pdf>.

National Research Council (NRC). 1997. U.S. Supersonic Commercial Aircraft: Assessing NASA's High Speed Research Program. Washington, D.C.: National Academy Press. Available online at <www.nap.edu/catalog/5848.html>.

NRC. 2000. Uninhabited Air Vehicles: Enabling Science for Military Systems. Washington, D.C.: National Academy Press. Available online at <www.nap.edu/catalog/9878.html>.

NRC. 2001. Commercial Supersonic Technology: The Way Ahead. Washington, D.C.: National Academy Press. Available online at <www.nap.edu/catalog/10283.html>.

NRC. 2002a. For Greener Skies: Reducing Environmental Impacts of Aviation. Washington, D.C.: National Academy Press. Available online at <www.nap.edu/catalog/10353.html>.

NRC. 2002b. Future Flight: A Review of the Small Aircraft Transportation System Concept. Transportation Research Board. Washington, D.C.: The National Academies Press. Available online at <www.nap.edu/catalog/10319.html>.

NRC. 2002c. Implications of Emerging Micro and Nano Technologies. Washington, D.C.: The National Academies Press. Available online at <www.nap.edu/catalog/10582.html>.

NRC. 2002d. Small Wonders, Endless Frontiers: A Review of the National Nanotechnology Initiative. Washington, D.C.: The National Academies Press. Available online at <www.nap.edu/catalog/10395.html>.

NRC. 2003. An Assessment of NASA's Aeronautics Technology Programs. Washington, D.C.: The National Academies Press. Available online at <www.national-academies.org/aseb>.

Pfenninger, W. 1987. Long-range LFC [laminar flow control] Transport. Research in Natural Laminar Flow and Laminar-flow Control. J. Hefner and F. Sabo, compilers. NASA CP-2487, Part 1, pp. 89-115.

Roco, M.C. 2002. National Nanotechnology Initiative and a Global Perspective. Presented at a National Science Foundation Symposium: Small Wonders, Exploring the Vast Potential of Nanoscience. Washington, D.C., March 19. Available online at <http://www.nano.gov/smwonder_slide.pdf>.

5

Process for Change

As discussed in previous chapters, a national vision, clear technology goals, well-defined organizational roles, and strong, focused leadership are necessary to improve the national and international competitiveness of the U.S. aeronautics industry and enable the air transportation system to satisfy increased demands for air travel without degrading system safety, security, environmental compatibility, or consumer satisfaction.

The role of the federal government in air transportation, as in other modes of transportation, is played by various agencies with sometimes conflicting agendas; these agencies must deal with (1) an intensively competitive carrier industry that rarely earns the cost of capital and (2) public infrastructure (airports, in the case of aviation) that makes up a major part of the physical system and is provided by other units of government (state and local) that also compete among themselves for service and investment to meet their own needs.

The aviation system *is* unique in that it has one federal agency (NASA) responsible for long-range research and development and another agency (FAA) that supplies traffic management systems and services and regulates the carriers and manufacturers. The cultures, missions, and operating practices of NASA's aeronautics enterprise and the FAA are quite distinct, as would be expected when comparing a research organization with an operational organization. Nonetheless, they are the federal government's principal agents for operating and improving the technical capabilities of the air transportation system.

One role of government is to support research in areas related to the public good, such as aviation safety, security, environmental effects, and other areas where the performance of the air transportation system impacts society. The purpose of governmental involvement is to bring advanced concepts and technologies to the point where private investment can be justified by industry. Government can make its involvement more effective by supporting and participating in noncompetitive research collaborations (including international collaborations) related to aviation safety and environmental effects (e.g., collecting data and developing models).

Using a flexible approach to government-industry relationships, the federal government can also support precompetitive research by U.S. industry, where it can cost effectively advance the current state of research and technology in ways that are most likely to make the transition from the research laboratory to commercial development. This transition is often difficult because the technology goals of NASA research programs often leave technology in a state that industry considers too immature to justify commercial development. The federal government and the aeronautics industry both operate in cost-constrained environments that encourage managers to rely on other organizations to fund research whenever possible. Strong, inspired leadership—by government and industry—will be needed to overcome these problems.

Strong leadership will also be required to ensure that research and technology are planned and conducted in the context of a well-organized and broadly supported process that has a comprehensive strategy for overcoming key challenges and can achieve the national vision for commercial aviation. Research should be guided by a consistent set of system performance requirements, operational concepts, system architectures, and implementation plans. Any other approach, even if it produces breakthrough technologies in selected areas, is likely to have a difficult time making the systemic improvements that will be necessary to keep pace with the long-term growth in demand for air transportation.

The importance of establishing strong interagency leadership that establishes a long-term vision and goals, coordinates interagency research, and conducts periodic reviews of the national aeronautical research and development programs is highlighted in the final report of the Commission on the Future of the U.S. Aerospace Industry (2002). The federal government has also established a joint aviation system program office, motivated in part by the perception that

- The demand for air transportation will exceed planned capacity improvements.
- A strategic realignment of government resources is needed to enhance mobility and improve the benefits provided by aviation research.
- Government leadership is needed to develop a unified national plan.

The joint program office will be guided by a policy committee chaired by the secretary of transportation and including the FAA and NASA administrators and senior executives from the Departments of Commerce, Defense, and Homeland Security.[1]

Developing a public-private consensus on a long-term vision and goals will be complicated by the different concerns of different stakeholders. Especially in times of financial difficulty, airlines understandably are highly cost sensitive and have a hard time looking past the immediate future. In addition, the FAA is forced by the nature of its close supervision by Congress, its own technical limitations, and intense pressure from the airlines to be conservative in the introduction of new technologies.

Many so-called scientific and engineering breakthroughs are the result of discoveries made 10 to 20 years earlier. Success often requires persistence, the willingness to challenge conventional wisdom, and/or a change in circumstances that significantly alters what is possible and practical. For example, the idea of the gas turbine reportedly is described in a British patent granted in 1791 (Moss, 1944), but use of the gas turbine for aircraft propulsion proved to be an elusive goal. As late as 1924, an investigator for the U.S. Bureau of Standards concluded that "jet propulsion would be impractical for either civilian or military purposes: The top speed of a jet-powered aircraft would be only 250 miles per hour and fuel consumption would be four times higher than piston engines" (Mandeles, 1998). Research continued nonetheless, but the first flight of a gas turbine propulsion system did not take place until 1941, when the imperative of war spurred massive aviation research and technological advances in materials and other fields made this achievement possible. World War II also laid the foundation for a greatly expanded air transportation system by training thousands of pilots, creating a huge inventory of surplus military aircraft and airports, and producing many other advances in the state of the art of aviation technology.

[1]John Kern, Federal Aviation Administration, briefing to Alan Angleman, National Research Council, on April 11, 2003.

Action necessary for securing the future for the U.S. air transportation system is encapsulated in the process for change that is defined in the following summary recommendation:

Recommendation 5-1. Process for Change. Establish air transportation as a national priority with strong, focused leadership. Air transportation system technology planning and development should be done in the context of a process driven by the needs of system users and the nation as a whole.

1. Implement a public/private process for change, as follows:
   - Designate a federal agency or office to provide strong leadership in overcoming the challenges faced by the U.S. air transportation system.
   - Establish an interagency process for developing and achieving a widely endorsed long-term vision of the air transportation system that includes a clear set of guiding principles and a strategy for overcoming transitional issues.
   - Document the process.
   - Coordinate action and resolve disputes among stakeholders in the aviation community with different concerns and priorities (e.g., manufacturers and operators; executives and employees; pilots, controllers, and passengers; local, federal, and state governments; regulators; the military; and general aviation).
   - Gather and analyze feedback on how well the process is working from the perspective of all interested parties, especially when conditions change, to identify problems before serious incidents or disruptions occur and to recognize new opportunities.
   - Formally review the process and process outputs at least every 4 years.
   - Update the process.
2. The output of the process should include the following:
   - A better understanding of future demand for air transportation to make sure that changing trends will be detected as soon as possible.
   - A unified, long-term national vision endorsed and supported by the aeronautics community as a whole and cognizant federal agencies.
   - Broad public policies to support the vision.
   - Long-term operational concepts to meet the vision and to serve as a continuing resource for guiding change and coordinating action by different parties.
   - System architectures to realize the operational concepts.
   - An understanding of how the U.S. air transportation system of the future will fit into the national (intermodal) transportation system and international air transportation system.
   - Validated research and technology requirements.

- An implementation plan to achieve all of the above, including a clear understanding of government and industry roles in developing precompetitive and noncompetitive aeronautical research and transitioning the results of civil and military government research to commercial development.

3. A comprehensive suite of system models should be developed, validated, and maintained to support informed decision making throughout the process. Models should encompass the following:
   - demand
   - economics
   - environmental effects
   - existing and new technologies
   - human performance
   - interactions with other modes of transportation
   - new operational concepts
   - organizational factors
   - security threats and preventive measures
   - system engineering
   - transition (from old to new technologies, systems, and organizational structures)
4. A commitment should be made to support a stable long-term research program to provide the knowledge, tools, and technologies needed throughout the process. At a low level, the research program should investigate innovative research ideas that challenge accepted precepts.

The Commission on the Future of the U.S. Aerospace Industry issued a report in 2002 with recommendations for federal action to ensure that the United States would maintain a robust aerospace industry in the 21st century. The scope of the Aerospace Commission's report is much broader than that of this report, and the Vision 2050 Committee was not chartered to validate the results of the Aerospace Commission. However, many of the findings and recommendations in this report are supported by the Aerospace Commission's recommendations:

> Recommendation #1. The integral role aerospace plays in our economy, our security, our mobility, and our values makes global leadership in aviation . . . a national imperative. . . . The Commission, therefore, recommends that the United States boldly pioneer new frontiers in aerospace technology. . . .

> Recommendation #2. The Commission recommends transformation of the U.S. air transportation system as a national priority. . . .

> Recommendation #9. . . . basic aerospace research . . . enhances U.S. national security, enables breakthrough capabilities, and fosters an efficient, secure and safe aerospace transportation system (Commission on the Future of the U.S. Aerospace Industry, 2002).

A final word on the current state of the air transportation industry in the United States. As recently as the summer of 2001, many travelers were dreading air transportation because of extensive delays associated with undercapacity of the system. That all changed on 9/11. Demand for air transportation has not yet returned to peak levels. Most U.S. airlines continue to struggle for survival, and some have filed for bankruptcy. The situation undermines the argument that strong action is urgently needed to avert a crisis of undercapacity in the air transportation system. Yet that remains the case. History shows that crises of confidence or international conflict can depress the demand for air transportation, but only over the short term. In every earlier situation, the long-term trend of increasing demand has reasserted itself. Assuming that current events have fundamentally and permanently changed public demand for transportation by air is not a sound basis for planning the long-term future of the air transportation system. Current events have provided an opportunity to get ahead of the problem; hopefully government and industry will be able to make the most of this opportunity.

REFERENCES

Commission on the Future of the U.S. Aerospace Industry. 2002. Final report of the Commission on the Future of the U.S. Aerospace Industry. Washington, D.C.: U.S. Department of Commerce, International Trade Administration, Office of Aerospace. Available online at <www.aerospacecommission.gov/AeroCommissionFinalReport.pdf>.

Mandeles, M. 1998. The Development of the B-52 and Jet Propulsion. Maxwell Air Force Base, Al.: Air University Press. Available online at <www.maxwell.af.mil/au/aul/aupress/Books/Mandeles_B52/PDF/Mandeles.pdf>.

Moss, S., 1944. Gas turbines and superchargers. Transactions of the American Society of Mechanical Engineers 66:351–371. Cited in <http://www.asme.org/history/brochures/h100.pdf>.

Findings, Recommendations, and the Big Question

Given below is a complete list of the committee's findings, recommendations, and the big question, in the order in which they appear in the report. The bulleted items in recommendations containing lists of research areas are either listed alphabetically or grouped topically. The committee did not prioritize research areas in the lists.

Recommendation 1-1. Goals. The future vision for the air transportation system should be supported by research and technology goals leading to improved performance. Measurable long-term targets supported by sound analyses should be established to assess progress toward the goals. Research should support the establishment of quantifiable goals in areas where progress is difficult to measure.

Finding 1-1. Challenge of Increased Demand. The continued success of aviation and the benefits that it provides will require changes to accommodate increased demand. This is the most critical long-term issue facing all aspects of the air transportation system. Issues associated with safety and security, capacity, environmental compatibility, and consumer satisfaction are all exacerbated by greater demand, and the effectiveness of near-term solutions in each of these areas will be diminished by long-term growth in demand for air transportation in the United States.

Finding 1-2. Going Beyond Business as Usual. Business as usual, in the form of continued, evolutionary improvements to existing technologies, aircraft, air traffic control systems, and operational concepts, is unlikely to meet the needs for air transportation that will emerge over the next 25 to 50 years. The likely result would be an air transportation system where growth in demand has been greatly curtailed by undercapacity in the air traffic management system; the environmental effects of aviation; customer dissatisfaction with available levels of comfort, convenience, and cost; and/or factors related to safety and security.

The Big Question. How can change within the air transportation system be accelerated quickly enough and directed with enough agility to avoid problems and achieve future goals while managing (1) the influence of increased demand and other external pressures and (2) conflicts between different goals and stakeholders? How can the system be prevented from changing too slowly, drifting, or going in the wrong direction?

Finding 1-3. Context for Future Requirements. Valid research requirements for the air transportation system depend on understanding how the U.S. air transportation system of the future will fit into both the national (intermodal) and international air transportation systems.

Recommendation 1-2. National Vision. The process of improving the long-term performance of the air transportation system—and organizing a corresponding long-term research and technology program—should start with a unified, widely endorsed, national vision that specifies goals in each key area of interest to the commercial aviation community. The vision should establish goals related to safety and security, the capacity of the air transportation system, environmental compatibility (noise and emissions), the satisfaction of consumer needs, and industrial competitiveness. It should include a clear set of guiding principles and a strategy for overcoming transitional issues.

Recommendation 1-3. Leadership. No single organization has the responsibility and authority for developing a comprehensive solution to the challenges faced by the U.S. air transportation system. Strong, focused leadership is needed. Federal leadership should be exercised by an agency or office with (1) the responsibility, authority, and financial resources necessary for defining air transportation system architectures through a centralized planning function, (2) an understanding of the interactions among system performance parameters, demand, and economic factors, such as the methods used to fund federal activities in support of the air transportation system, and (3) the credibility and objectivity to garner the active support of other air transportation stakeholders in government, industry, and the general public. This requires, among other things, a leadership group composed of individuals with a broad aviation perspective and a willingness to accept the risks of (1) looking ahead and (2) allowing others to help define the future.

Finding 2-1. The Challenge. Developing meaningful and useful operational concepts stemming from a broadly defined vision of the air transportation system 25 to 50 years hence is a critically important task in the process of improving the performance of the system.

Recommendation 2-1. Operational Concepts 2050. The federal government, working with other stakeholders in the air transportation system, should develop a coherent set of operational concepts to support a vision for the air transportation system in 2050 to guide (1) long-term research and (2) the evolution of and transition to a more advanced air traffic management system. The set of operational concepts should be continually, objectively, and rigorously evaluated (for example, through comprehensive simulation and modeling) and iterated to reflect feedback from stakeholders, conflicts between alternative concepts, and the best understanding of the future costs, benefits, and requirements that are likely to evolve in response to changes in the real world, the current state of technology and systems operations, and future expectations. Strong national leadership should coordinate the efforts of all involved federal agencies and other stakeholders in the air transportation system to build toward concepts that best support the vision.

Recommendation 2-2. Enabling Technologies. Enabling technologies applicable to a wide range of operational concepts should be developed in parallel with development and evaluation of long-term operational concepts so that the necessary technologies will be ready for whichever operational concept proves to be most beneficial. Technology areas of particular interest include the following:[1]

- Automation technologies applicable to fully automated systems; automated decision aids; and information systems for communication, visualization, situation assessment, and the prediction of future conditions.
- Technologies that support distributed, collaborative decision making and that foster coordination and interactions among multiple human and automated elements of the system.
- Methods and technologies for moderating and abating the impact of noise and emissions locally, regionally, and globally.
- Methods and technologies for predicting or directly sensing the magnitude, duration, and location of wake vortices and the potential to reduce separation standards without compromising safety.
- Methods for identifying (1) the information required for situation awareness when humans are assigned novel (untried) tasks in future operational concepts and (2) sensor, computing, and display technologies for better supporting situation awareness, judgment, decision making, and planning. Relevant technologies include synthetic vision, cockpit and controller displays for novel air traffic management functions, fast-time simulation and computational functions for predicting future conditions, and alerting. These methods and technologies should be investigated for their potential to (1) reduce separation standards without compromising safety and (2) enable changes in the roles of humans within the system.
- Systems-engineering methods that are (1) capable of conceiving and analyzing systems of the complexity of air transportation and (2) suitable for governing the design, testing, and implementation of these systems.
- Avionics technologies that will provide ubiquitous and transparent communication, navigation, and surveillance capabilities; enable cost-effective, reliable air traffic management; and contribute to the reduction of separation standards without compromising safety.

Recommendation 2-3. Design of Complex Human-Integrated Systems. The design of human-integrated systems—that is, systems that rely on the combined activities of humans and machines—presents significant challenges at every level, from the systems level (e.g. creating effective teamwork within operations involving many human operators and automated system elements) to the detailed design level (e.g., developing operating procedures and system displays). Research in the following areas is required to understand and address these challenges:

- A broad, interdisciplinary approach that includes technology designers, users, and experts in human and organizational performance from the earliest stages of conceptual design through final implementation to develop technology that effectively supports human behavior and recognizes the need for concurrent design of procedures, training, and technology.
- Geographically distributed activities, such as coordinated decision making and planning, that are mediated by computers and automated system elements.
- Human factors, human-automation interactions, and functioning of teams of humans and automated system elements.

[1]In this and other recommendations that list research areas, the bulleted items are either listed alphabetically or grouped topically. The committee did not prioritize research areas within each list, and bulleted items are not listed by priority.

- Specific impact of newly automated functions and changes in human roles.
- System engineering methods for addressing organizational and systemwide issues.

Finding 2-2. Nontechnological Impediments to Success. Technological research alone is insufficient to achieve the future vision. Research is also needed to (1) better understand the economic, environmental, political, institutional, and managerial factors involved in achieving key goals, (2) take advantage of synergies among these factors, and (3) overcome related impediments.

Recommendation 2-4. Research Needs Beyond Technology Development. The federal government should also support research to develop improved processes and methods in the following nontechnology areas:

- Assessment of economic factors, such as taxes, fees, and subsidies established by the government, that influence (1) the demand for and the supply of air transportation services and (2) key decisions made by organizations and individuals involved in the provision and use of the air transportation system.
- Modification of regulations, certification requirements, and operating procedures.
- Prediction and resolution of conflicting objectives of different stakeholders in the air transportation system.
- Understanding societal concerns about aircraft noise and emissions.

Recommendation 3-1. Value of Modeling and Simulation. Federal agencies involved in modeling and simulation of the air transportation system should make complementary use of field tests, laboratory tests, modeling, analysis, and simulation to improve their ability to (1) measure systemwide behavior of the air transportation system, (2) assess the performance of proposed operational concepts, technologies, and other changes, and (3) make informed investment decisions that reduce the schedule, cost, and technical risk of system improvements.

Recommendation 3-2. Management of System Models. Federal agencies that support research in aviation system models should improve their coordination, especially with regard to the following:

- Ensuring that the federal investment for research and development in aviation models focuses on key issues, avoids unnecessary duplication, and encourages cooperation among developers.
- Encouraging participation of industry and academia in modeling and simulation research and development relevant to government needs.
- Establishing widely accepted criteria for the maintenance and validation of models.
- Identifying models that are most important to government policy decisions.
- Making those models more widely available to users inside and outside government.
- Ensuring that modeling and simulation results are used appropriately by decision makers involved in developing the future aviation system.

Recommendation 3-3. System Modeling Research. The government and other interested parties should support additional research in the following critical areas:

- Improving the interoperability of high-fidelity, detailed, data-intensive, long-run-time models of the U.S. air transportation system and the higher-level fast-time, abstract models necessary to analyze overall system performance under a variety of different assumptions so that both types of models can be brought to bear on relevant problems. (It may be feasible to develop models with adjustable resolution that can simplify variables for faster run time when those variables are critical to the analysis being performed.)
- Modeling and simulation methods suitable for safety analysis, which inherently require a detailed level of modeling that includes all the factors that contribute to safety, including human performance and sociotechnical aspects of the system. (Additional fundamental research and development is required before these methods can enter widespread use. New approaches should be pursued using systems theory as well as new nontraditional chain-of-event models.)
- Modeling demand and demand allocation for air transportation services, particularly as it relates to airline schedule changes, including city-pairs, routes (including altitudes and way points), time of day, and the establishment (or elimination) of hub airports. (Dynamic interactions between changing or radically new operational concepts and technologies and user behavior, as they relate to all modes of transportation and other factors external to the air transportation system, must be better understood to ensure the right problems are being addressed.)
- Requirements, methods, and standards for validating individual models and suites of models.
- Understanding how to connect models to form a suite of system models that includes nonlinear dynamic interactions and emergent properties.
- Understanding the role of humans in the aviation system of the future and how to communicate this understanding in a convincing and supportable way. (Including computational human performance models in current simulations and using human-in-the-loop simulations is critical.)

Finding 4-1. Advanced Aircraft Technology. Improvements in aircraft performance are critical to achieving necessary improvements in almost every aspect of the overall performance of the air transportation system. Innovative long-range research leading to the implementation of new operational concepts is also required for the air transportation system to take full advantage of gains in the performance of commercial aircraft.

Recommendation 4-1. Aircraft Research and Technology. To improve the performance of aircraft through 2025, federal agencies should continue to support research leading to evolutionary improvements in aircraft performance. Looking out to 2050, however, research should support innovative concepts aimed at major advances in performance. In addition, agencies should continue to monitor research in related emerging technologies, such as nanotechnologies, and support research aimed at aircraft applications as emerging technologies mature. The areas listed below are prime candidates for this kind of long-term research.

- Analytical tools, advanced technologies, and the fundamental science behind both, to reduce the need for costly hardware testing and to more easily achieve overall research goals (especially in emerging technologies).
- Composite materials with improved qualities that would increase their use in airplane structures and reduce aircraft weight.
- Environmental consequences of aircraft noise and emissions locally, regionally, and globally, to better understand those consequences and support the establishment of better informed priorities and goals for noise and emissions reduction that (1) reflect the need for integrated approaches (involving advances in airframes, engines, and operational procedures) to meet environmental goals and (2) accurately account for the tradeoffs among different environmental goals and different approaches to achieving those goals.
- Low emissions combustor technology, to (1) reduce substantially emissions of NO*x* and particulate matter at airports (to improve local and regional air quality) and at cruise altitudes (to reduce global climate effects) and (2) reduce emissions produced by engines with high pressure ratios.
- Nanotechnology, to explore its long-range potential for dramatically enhancing aircraft performance through the development of advanced avionics (computing, sensors, and active distributed controls) and high-performance materials.
- Nontraditional aircraft configurations, including but not limited to (1) the blended-wing-body and (2) the strut-braced or joined wing, to improve aircraft productivity and efficiency and reduce noise and emissions.
- Passive and active control of laminar and turbulent flow on aircraft wings (laminar flow to increase cruise efficiency and turbulent flow to increase lift during take-off).
- Nontraditional power and propulsion concepts and technologies, especially concepts and technologies that support the use of alternative fuels, such as fuel cells (which may have application as auxiliary power units in the foreseeable future) and high-density storage of hydrogen to improve the feasibility of using it as a propulsion fuel.
- High-temperature engine materials and advanced turbomachinery, including (1) lower speed, highly loaded, fan drive turbines and fan reduction gears; (2) very large fan systems, which require advances in manufacturing and material systems; (3) boundary layer control on turbomachinery airfoils, to improve component efficiency and packaging; (4) aspirated turbomachinery components, which could greatly reduce noise and improve component efficiency; and (5) other innovative concepts, such as interturbine burning or volume cooling ahead of or in the compressor by means of a mist of water or other coolant.

Recommendation 5-1. Process for Change. Establish air transportation as a national priority with strong, focused leadership. Air transportation system technology planning and development should be done in the context of a process driven by the needs of system users and the nation as a whole:

1. Implement a public/private process for change, as follows:
   - Designate a federal agency or office to provide strong leadership in overcoming the challenges faced by the U.S. air transportation system.
   - Establish an interagency process for developing and achieving a widely endorsed long-term vision of the air transportation system that includes a clear set of guiding principles and a strategy for overcoming transitional issues.
   - Document the process.
   - Coordinate action and resolve disputes among stakeholders in the aviation community with different concerns and priorities (e.g., manufacturers and operators; executives and employees; pilots, controllers, and passengers; local, federal, and state governments; regulators; the military; and general aviation).
   - Gather and analyze feedback on how well the process is working from the perspective of all interested parties, especially when conditions change, to identify problems before serious incidents or disruptions occur and to recognize new opportunities.
   - Formally review the process and process outputs at least every 4 years.
   - Update the process.
2. The output of the process should include the following:
   - A better understanding of future demand for air transportation to make sure that changing trends will be detected as soon as possible.

- A unified long-term national vision endorsed and supported by the aeronautics community as a whole and cognizant federal agencies.
- Broad public policies to support the vision.
- Long-term operational concepts to meet the vision and to serve as a continuing resource for guiding change and coordinating action by different parties.
- System architectures to realize the operational concepts.
- An understanding of how the U.S. air transportation system of the future will fit into the national (intermodal) transportation system and international air transportation system.
- Validated research and technology requirements.
- An implementation plan to achieve all of the above, including a clear understanding of government and industry roles in developing precompetitive and noncompetitive aeronautical research and transitioning the results of civil and military government research to commercial development.

3. A comprehensive suite of system models should be developed, validated, and maintained to support informed decision making throughout the process. Models should encompass the following:
   - demand
   - economics
   - environmental effects
   - existing and new technologies
   - human performance
   - interactions with other modes of transportation
   - new operational concepts
   - organizational factors
   - security threats and preventive measures
   - system engineering
   - transition (from old to new technologies, systems, and organizational structures)
4. A commitment should be made to support a stable long-term research program to provide the knowledge, tools, and technologies needed throughout the process. At a low level, the research program should investigate innovative research ideas that challenge accepted precepts.

Appendixes

A

Statement of Task and Study Approach

PHASE 1 STATEMENT OF TASK. ASSESSING THE VISION

Initially, the committee will assess the visions and goals contained in the documents listed below as they pertain to civil aviation in the United States.[1]

- National Aeronautics and Space Administration (NASA), "Goals and Objectives for the Aerospace Technology Enterprise," 1997 (revised 2001), available online at <www.aerospace.nasa.gov/goals/index.htm>
- National Science and Technology Council, *National Research and Development Plan for Aviation Safety, Security, Efficiency, and Environmental Compatibility*, 1999, available online at <www.volpe.dot.gov/infosrc/strtplns/nstc/aviatrd/index.html>
- Federal Transportation Advisory Group, *Vision 2050: An Integrated Transportation System*, 2001, available online at <http://scitech.dot.gov/polplan/vision2050/index.html>
- The related white paper "Next Generation Air Transportation System," Aerospace Transportation Advisory Group, 2001, available from the Aeronautics and Space Engineering Board
- NASA, *Aeronautics Blueprint*, 2002, available online at <www.aerospace.nasa.gov/aero_blueprint/index.html>

The first document in the above list proposes key goals for national civil aeronautics research covering both near- and far-term applications, along with program initiatives intended to achieve those goals. The second report recommends a new national transportation vision, and the last three identify current constraints and recommend research and development investments to improve the air transportation system over the next 25 years, respectively.

The committee's assessment will identify compatibilities and incompatibilities among the visions and goals described in the above documents. The committee will also hold a workshop to solicit inputs from the aeronautics community regarding the extent to which advanced technology will be able to achieve future goals and visions in the next 25 to 50 years. The committee's assessment will also consider how advanced technology can help civil aviation succeed in the new threat and heightened security environment that exists in the aftermath of the attacks of September 11, 2001. The committee will refrain from creating a new vision.

Phase 1 will result in two reports. The first report will contain the presentations of the workshop (with no findings or recommendations). The workshop participants will be asked to provide electronic copies of their presentations, and the workshop report may be disseminated in the form of a CD-ROM computer disk. No later than August 2002, the committee will also issue a letter report summarizing its assessment of future goals and visions for national civil aviation. This report was requested by the White House Office of Science and Technology Policy.

PHASE 2 STATEMENT OF TASK. ASSESSING KEY TECHNOLOGIES

Phase 2 will assess technology goals for 2050. Specific tasks are as follows:

- Identify the extent to which expected advances in key technologies could achieve the aviation vision in 2025 and 2050.
- Identify key technological goals that will not be met by

[1]Given the long-term nature of this study, it did not assess near-term plans for improving civil aviation, such as the Federal Aviation Administration's Operational Evolution Plan.

continued evolution of existing technologies and programs.

- Identify critical research initiatives needed to reach key goals.
- Determine if major changes in national aeronautical research and development programs would make it easier to achieve the key goals.

The committee's third and final report, to be issued 26 months from study initiation, will summarize the results of Phase 2, including findings and recommendations for action.

The committee will not collect classified military information. The committee will also avoid making findings or recommendations in areas that are discussed in the source documents for this study, but which are outside the scope of this study. Out-of-scope topics include development of technology for building exoatmospheric flight vehicles; the organization and role of government agencies and advisory groups; levels of government funding (the committee's recommendations should focus on funding priorities rather than any particular level of funding); methods of interaction among government, industry, and academia; sources and use of private capital; size of the required workforce; legal and regulatory frameworks; and U.S. defense, social, foreign, and economic policies. Satellite-based communications, navigation, and surveillance systems are within the scope of the study.

STUDY APPROACH

During Phase 1, as requested by the study sponsors, the committee refrained from creating a detailed vision of its own. Instead, the committee evaluated existing visions and noted shortcomings that should be corrected in future visions.

During Phase 2, the assessment of technology goals for the year 2050 began with the four tasks outlined above. Upon review and discussion—and with the support of the sponsors—the committee agreed to focus its efforts on three technology areas where research in the short term could lead to substantial long-term improvements in key areas of the vision and goals:

- system modeling and simulation
- performance of the air transportation system
- aircraft performance

B

Comparative Assessment of Goals and Visions

In response to the statement of task for Phase 1 of this study, the Committee on Aeronautics Research and Technology for Vision 2050 assessed the future visions and goals for U.S. civil aviation as expressed in five source documents (see Appendix A). To gain additional insight into U.S. visions and goals, the committee also included in its assessment a comparable vision for civil aeronautics in Europe, *European Aeronautics: A Vision for 2020—Meeting Society's Needs and Winning Global Leadership*, which was produced by the Group of Personalities and released in 2001 (available online at <http://europa.eu.int/comm/research/growth/aeronautics2020/en/>).

NASA'S NEW GOALS

After the committee began its work, NASA adopted new goals and objectives for its aeronautical research program. These changes were part of a reformulation of NASA's agency-wide vision, mission, and objectives, which are now as follows:

- NASA's vision
  —to improve life here
  —to extend life to there
  —to find life beyond
- The NASA mission
  —to understand and protect our home planet
  —to explore the universe and search for life
  —to inspire the next generation of explorers
  . . . as only NASA can
- Role of the Aerospace Technology Enterprise
  —to pioneer and validate high-payoff technologies: to improve the quality of life, to enable exploration and discovery, and to extend the benefits of innovation throughout our society

Based on the above, NASA's Aerospace Technology Enterprise established Aeronautics Technology Strategic Theme Objectives, which are listed in the second column of Table B-1. NASA's old goals and objectives are listed in the first column.

In making these changes, NASA has replaced time-phased, quantitative goals (e.g., "double the aviation system capacity within 10 years, and triple it within 25 years") with open-ended, qualitative goals (e.g., "enable more people and goods to travel faster and farther, anywhere, anytime, with fewer delays."). For some technologies quantifiable goals are difficult to define, and limiting research to areas with easily quantifiable goals would reduce the scope of NASA's research to a subset of the overall problem. On the other hand, quantifiable goals could be readily established in many areas, and even as NASA moves away from quantitative research goals, the report of the Commission on the Future of the U.S. Aerospace Industry (2002) concludes that "the Administration and Congress should adopt . . . aerospace technology demonstration goals for 2010 as a national priority." In the area of air transportation, the specific goals endorsed by the Aerospace Commission are as follows:

- Demonstrate an automated and integrated air transportation capability that would triple capacity by 2025.
- Reduce aviation fatal accident rate by 90 percent.
- Reduce transit time between any two points on earth by 50 percent.
- Reduce aviation noise and emissions by 90 percent.

The first three goals are equivalent to NASA's now obsolete 25-year goals for improving system capacity, safety, and mobility. The emission goal does not specify what emissions should be reduced. In any case, a 90 percent reduction is more ambitious that NASA's previous 25-year goals, which were to reduce emissions of oxides of nitrogen (NO_x) and

TABLE B-1 Comparison of Future Goals and Visions for Civil Aeronautics

NASA Goals and Objectives						
Superseded	**Current**	**National R&D Plan**	**2050 Vision**	**Next Generation Air Transportation System**	**NASA Aeronautics Blueprint**	**European Aeronautics: A Vision for 2020**
Safety • Accident rate reduced 90% in 25 years	**Safety and Security** • Protect air travelers and the public: Decrease the aircraft accident rate and mitigate the consequences of accidents • Efficiently decrease aviation system vulnerability to threats • Mitigate consequences of hostile acts	**Safety and security** • Risk management • Prevention of accidents and incidents • Reduction of injuries when accidents occur • Reduction of casualties, system disruption, and damage when incidents occur	**Safety and security** • No fatalities or injuries —Human-centered systems to compensate for human error —Education and training for operators and users (lifetime learning) —Accident prevention —Reduction of injuries when accidents occur —Prevention of incidents: reduction of casualties when incidents occur without disrupting normal operations	**Safety and security** • Decentralized air transportation system to improve safety in case of system failure, human error, or terrorist attack	**Safety and security** • Improved situational awareness in all weather • Aircraft health monitoring systems, self-healing systems, and adaptive fault-tolerant controls to respond to system failures and human error • Aircraft hardening to withstand explosions • Improved monitoring of projected flight paths to prevent aircraft collisions and detect unauthorized diversions • Prevention of malicious or ill-advised pilot actions from causing an accident • Automated passenger identification and threat assessment	**Safety** • Reduce accident rate by 80% • Aircraft systems that lighten the burdens on the crew, help them make correct deci sions, and reduce the impact of human error • Higher standards of training • Monitoring systems that react to technical problems a they occur
Mobility • Reduce intercity door-to-door transportation time by 67% in 25 years • Reduce long-haul transcontinental travel time by 50% in 25 years **Capacity** • Triple the capacity of the air transportation system in 25 years **Technology innovation** • Revolutionary technologies to enable fundamentally new system capabilities **Engineering innovation** • Advanced tools, processes and culture to enable rapid, high-confidence, and cost-efficient design of revolutionary systems	**Mobility** • Enable more people and goods to travel faster and farther, anywhere, anytime, with fewer delays **New aerospace missions** • Pioneer novel aerospace concepts to support earth and space science missions	**National Airspace System** • Improve system capacity, capabilities, cost-effectiveness, and services • Architecture definition and evolution • Architecture implementation • Improving air traffic operations —Involving air traffic controllers in developing new systems and associated training —Human factors research for controller operations, system maintenance, and improved weather services • Developing breakthrough technologies to meet growing demand	**Efficiency** • Integrated • Intermodal **Advanced technology** • Improved transportation system definition tools and vehicle design tools and methods • Optimized vehicle and system operations • New classes of superefficient, intelligent, reliable, and environmentally friendly vehicles	**Efficiency** • Automated management of the airports and airspace for all types of aircraft, day and night, in all weather conditions • Cultural and institutional changes • Radically new air transportation management systems • System definition tools	**National Airspace System** • Weather —Reduced impact of low visibility —Better observation and prediction of adverse weather and vortices • Traffic optimization —Automated, distributed air traffic management • High-capacity airports —Integrated arrival, departure, and surface decision-support tools —Synthetic vision systems for aircraft —Airport design and operation models —Smart nontowered airports • CNS —Satellite-based CNS systems —Active and passive precision navigation/surveillance systems **Revolutionary vehicles** • Global range • Supersonic speed • Vertical lift and extremely short takeoff and landing • Long-duration uninhabited aircraft • Nanotechnology, variable aerodynamic shapes, and advanced propulsion and power systems **Educated workforce** • Motivating the next generation to work in aviation • Lifetime learning for the existing workforce • Multidisciplinary research using virtual laboratories	**Air transportation system** • System capacity increase 200% in all weather conditio • Sophisticated ground- an satellite-based CNS systems • Free Flight • Airports freed of noise-related operating restrictions • Air transportation integrated into an efficient multimod transportation system • Integrated ATM system that is so effective it become the de facto world standard **Educational policies to provide skilled aeronautics workforce** **Aircraft design and production** • Integrated design, manufacturing, and maintenance systems • Large-capacity aircraft (~1,200 passengers) • Supersonic speed • Innovative vertical takeoff and landing

Continued

TABLE B-1 Comparison of Future Goals and Visions for Civil Aeronautics (continued)

NASA Goals and Objectives: Superseded	NASA Goals and Objectives: Current	National R&D Plan	2050 Vision	Next Generation Air Transportation System	NASA Aeronautics Blueprint	European Aeronautics: A Vision for 2020
Noise • Perceived noise of new aircraft reduced 75% in 25 years **Emissions** • NO_x of new aircraft reduced 80% in 25 years • CO_2 of new aircraft reduced 50% in 25 years	**Environment** • Protect local environmental quality and the global climate by reducing aircraft noise and emissions	**Noise** • Reduce noise by 5 to 10 dB by 2007 and up to 20 dB by 2022 **Emissions** • Local effects: Reduce emissions of NO_x by 70% by 2005 and 80% by 2022 • Global effects: reduce emissions of CO_2 by 25% by 2007 and 50% by 2022	**Compatibility with the environment** • Noise • Land-management techniques • Noxious emissions and greenhouse gases —Energy-efficient air transportation system —Non-carbon-based fuels	**Compatibility with the environment**	**Noise** • Reduce noise restrictions on aircraft operations • Eliminate the need to soundproof homes near airports • Computational simulation of airflow in the engine and exhaust • Low-speed fans and nozzles • Low-noise airframe designs (e.g., morphing structures) **Emissions** • Reductions in NO_x to improve local air quality • Reductions in CO_2 to reduce global effects	**Noise** • Elimination of noise as a nuisance outside airport boundaries through reduction of perceived noise by 50%, better land planning and use, and systematic use of noise reduction procedures **Emissions** • CO_2 and NO_x emissions per passenger kilometer reduced by 50% and 80%, respectively, for new aircraft • Lower fuel use through drag reduction (using conventional and novel shapes), fuel additives, and better airframe/engine integration
						Quality and affordability of air transportation • Choice of routes and schedules • 99% on-time departures and arrivals in all weather • Quick airport check-in • Comfortable aircraft accommodations • Reduced costs for operators, passengers, and freight
						Primacy of the European aeronautics industry • New framework for companies to work together • Synergies between civil and military research • New standards of quality and effectiveness • Time to market cut in half
Commercialize technology						
	Support national security: • Leverage NASA aeronautics technology investments in partnership with the Department of Defense to support national defense					
			Independent of foreign sources of energy	**Independent of foreign sources of energy**		

NOTE: CNS, communications, navigation, and surveillance; ATM, air traffic management.

carbon dioxide by 80 percent and 50 percent, respectively. In order to serve as a reliable guide for research and policy changes, goals for emissions should clearly state whether the goals are for average emissions or "characteristic emission" levels, which are about 15 percent higher than the level of NO_x measured during the landing-takeoff cycle specified by the International Civil Aviation Organization. Goals should also be established for cruise operations (to limit climate effects) and landing and takeoff cycles (to limit effects on local and regional air quality). Another possibility would be to establish emissions goals that take into account the operational efficiencies of individual airlines, requiring them to meet goals in terms of emissions per revenue-passenger-mile, for example.

ASSESSING THE GOALS AND VISIONS

The committee identified compatibilities and incompatibilities in the visions and goals above, as they relate to civil aeronautics. The results are summarized in Table B-1 and discussed below. Although the above documents encompass a variety of time periods, starting as early as the late 1990s and extending as far as 2050, the committee concluded that all of the documents consistently emphasize three main thrusts:

- safety and security
- capacity of the air transportation system
- environmental compatibility (noise and emissions)

The visions point to a comprehensive approach to safety and security that includes both prevention of accidents and incidents and mitigation of consequences (in terms of injuries, damage to equipment, and disruption of the air transportation system). Not unexpectedly, the most recent visions—NASA's *Aeronautics Blueprint* and the current agency goals and visions—place much more emphasis on security than the visions created before September 11, 2001. As funding for transportation security increases, the magnitude of related technology development efforts is also likely to increase. As with the other major thrusts, long-term plans for developing technology to improve security should be based on a systematic approach that assesses the specific problems that need to be addressed and targets technologies accordingly. For example, security systems should be designed to predict and adapt to future threats to ensure that we are not in a constant state of preparing to "fight the last war." This requirement is not included in the current U.S. visions.

The visions point to the need for a similarly comprehensive approach when it comes to the capacity of the air transportation system. Research and technology efforts are called for to realize improvements in four areas: (1) the performance of each of the primary elements of the air transportation system: aircraft, air traffic management systems, and airports; (2) the integration of the air transportation system with other modes of transportation; (3) supporting systems such as communications, navigation, and surveillance systems and weather observation and prediction systems; and (4) the design and development processes used to create new technologies and products.

Some of the documents quantify their goals for reducing the noise of new aircraft. NASA's *Blueprint* and the European aeronautics vision, however, specify the ultimate goal in terms of operational impact: Aircraft noise should be reduced to the point where it is no longer a nuisance outside airport boundaries and airports are freed from operational restrictions related to noise. In terms of emissions, most of the visions deal only with NO_x and CO_2. The National Research and Development Plan takes a more open-ended view—it covers all emissions that affect local air quality, global climate, or atmospheric ozone. This broader view is important to ensure that future aeronautics research adequately considers both new understandings about the threat of other emissions and the complex interrelationships among various strategies for dealing with environmental problems. For example, high-altitude emissions of water vapor may be a particularly important environmental consideration for new supersonic aircraft, and some approaches for reducing NO_x may increase emissions of particulates.

The European aeronautics vision includes two areas that are not highlighted in any of the U.S. visions:

- quality and affordability of air transportation
- global primacy of the aeronautics industry

By including quality and affordability issues, the European vision acknowledges the importance of structuring research and development programs so that they are focused on providing air transportation services that users want to buy and are able to afford. The original 1997 version of NASA's goals spoke of reducing the cost of air travel by 50 percent within 20 years. However, this goal fell into disfavor with Congress, which seemed to view the meeting of customer demands as an industry responsibility that was inappropriate as a topic of NASA research. Congress reduced NASA's aeronautics budget to eliminate research related to this goal, so NASA eliminated the goal.

The European vision foresees the following future:

> In 2020, European aeronautics is the world's number one. Its companies . . . are winning more than 50% shares of world markets for aircraft, engines, and equipment. . . . The public sector plays an invaluable role in this success story. . . . Crucially, they are coordinating a highly effective European framework for research cooperation, while funding programmes that put the industry on more equal terms with its main rivals (Group of Personalities, 2001).

The future of the U.S. air transportation system is not necessarily tied to the future of the U.S. aeronautics manufacturing industry. Advanced aircraft and air traffic management systems could be procured from foreign suppliers if U.S. manufacturers fail to remain competitive. However, the supremacy of the U.S. aeronautics industry provides important national security and economic benefits. A U.S. aeronautics vision and research program that does not explicitly consider the importance of U.S. leadership in aeronautics could make it easier for the Europeans to achieve their vision of global leadership and market dominance.

One of the ways that the European vision foresees achieving global primacy is through greater cooperation and harmony among various elements of the aeronautics community throughout the European Union. Similarly, improved coordination and cooperation in this country would benefit the United States. Close cooperation is needed between NASA and the Federal Aviation Administration (FAA), for example, to establish requirements for NASA research that are relevant to the air traffic management systems the FAA is likely to procure in the future. The FAA's ability to support long-term systems analysis and requirements definition is limited, however, because so much effort is expended to solve more immediate problems and keep the air transportation system operating.

Commercialization of technology only shows up in NASA's goals and objectives. This is an important goal for a research agency like NASA, because the value of its aeronautics research is closely linked to its ability to transfer research results to other organizations that are directly involved in the development or production of aircraft, air traffic management systems, and other aviation products.

Two of the visions include independence from foreign sources of energy. The committee questions the wisdom of giving much consideration to this goal when formulating a national aeronautics research program because it does not believe that freeing one segment of the economy, such as air transportation, from foreign sources of energy would produce significant benefits if the economy as a whole remained dependent on foreign energy. Before diverting significant resources to research intended to free air transportation from foreign energy, the federal government should conduct a comprehensive, economy-wide assessment of various options for reducing U.S. dependence on foreign energy. Such a study might well conclude that the optimum strategy for reducing U.S. dependence on foreign energy should focus on nonaviation uses of petroleum products. A large reduction in the demand for jet fuel would not greatly reduce overall demand for petroleum products. In 2000, jet fuel accounted for less than 9 percent of U.S. consumption of petroleum products (EIA, 2002). In addition, design requirements are more stringent for aircraft engines, especially in terms of reliability and power density, than for virtually any other large-scale user of petroleum. New types of engines are unlikely to be adopted by the aviation industry unless they are first proven in other applications.

REFERENCES

Commission on the Future of the U.S. Aerospace Industry. 2002. Final Report of the Commission on the Future of the U.S. Aerospace Industry. Washington, D.C.: U.S. Department of Commerce, International Trade Administration, Office of Aerospace. Available online at <www.aerospacecommission.gov/AeroCommissionFinalReport.pdf>.

Energy Information Administration (EIA). 2002. Energy Consumption by Source, 1949-2000. Available online at <www.eia.doe.gov/emeu/aer/txt/tab0511.htm>.

Group of Personalities. 2001. European Aeronautics: A Vision for 2020—Meeting Society's Needs and Winning Global Leadership, p. 15. Available online at <http://europa.eu.int/comm/research/growth/aeronautics2020/en/>.

C

Biographies of Committee Members

RONALD R. FOGLEMAN is chairman and CEO, Durango Aerospace. He retired in 1997 as the Chief of Staff of the U.S. Air Force after 34 years of active duty. As a member of the Joint Chiefs of Staff, he served as a key military advisor to the Secretary of Defense, the National Security Council, and the President of the United States. With extensive operational and flight experience, General Fogleman held many senior command positions throughout his career, including Commander in Chief of the U.S. Transportation Command and Commander of the Air Mobility Command, in which he managed the operation of a complex air transportation system that interfaced with commercial airlines and the National Airspace System. After retirement, General Fogleman became president and COO of Durango Aerospace, Inc., an aviation consulting firm. He has chaired an Air Force Research Laboratory study on directed energy weapons for tactical platforms and has served as a member of NASA's Mars Program Independent Assessment Team, the congressionally mandated Commission to Assess United States National Security Space Management and Organization (the Rumsfeld Commission), and the Defense Policy Board.

JACK CLEMONS is the senior vice president of Lockheed Martin Transportation and Security Solutions in Rockville, Maryland. Mr. Clemons began his career at General Electric Corporation's Reentry Systems Division in Valley Forge, Pennsylvania. He worked on the NASA Apollo and Skylab programs for the TRW Systems Group in Houston, Texas, and on the NASA Space Shuttle program for IBM in Houston. Mr. Clemons joined Lockheed Martin's Air Traffic Management Company in 1992 as functional manager of software development and was director of en route programs, vice president for air traffic control engineering, and then senior vice president of engineering, technology and operations before taking his current assignment. Mr. Clemons graduated from the University of Florida with B.S. and M.S. degrees in aerospace engineering. Mr. Clemons contributed to the committee an industry perspective on the practical aspects of developing and fielding capacity, safety, and security enhancements to the U.S. air traffic control system.

WILLIAM B. COTTON is president of Flight Safety Technologies, Inc. Throughout the aviation industry, Captain Cotton is known as the "father of free flight," the air traffic management operating concept that is currently being developed and implemented to increase the safety, capacity, and operating efficiency in the nation's air traffic control system. His career includes 33 years with United Airlines, during which time he gained expertise in the areas of air traffic control and cockpit operating systems. While working for United, Captain Cotton held various positions, including manager of air traffic and flights systems and chairman of the board for Aeronautical Telecommunications Network Systems, Inc. He has a B.S. in aeronautical and astronautical engineering from the University of Illinois and an M.S. in aeronautical and astronautical engineering from the Massachusetts Institute of Technology.

EUGENE E. COVERT, NAE, is the T. Wilson Professor of Aeronautics, emeritus, at the Massachusetts Institute of Technology. His long and distinguished career in aerospace has spanned over 40 years in academia and has included additional stints as chief scientist of the U.S. Air Force, member and chairman of the Air Force Scientific Advisory Board, chairman of the Power and Propulsion panel of NATO's Advisory Group for Aerospace Research and Development, director of the Wright Brothers Facility, member or chair of numerous NRC study committees, and chairman of the NRC's Aeronautics and Space Engineering Board. Dr. Covert's experience provided an important perspective on trends in aeronautical research and development, particularly with regard to propulsion.

WILLARD J. DODDS has expertise in propulsion emissions technology and regulations. He is a consulting engineer for aircraft engine environmental issues at GE Aircraft Engines, one of two U.S. manufacturers of large jet engines. He is an expert in all aspects of aircraft engine combustion system design and development, including the design and development of high-performance and low-emission combustion systems. As such, he has an expert knowledge of engine emissions abatement technology and relevant regulatory considerations. He is currently chair of the International Coordinating Council of Aerospace Industries Association's Aircraft Noise and Engine Emissions Committee. In that capacity, he is the primary industry representative for interactions with the International Civil Aviation Organization on engine noise and emissions regulatory issues. Noise and emissions are long-term problems facing aviation, and Mr. Dodds helped the committee address this issue. He served on one other NRC committee.

WILLIAM W. HOOVER is currently a consultant for aviation, defense, and energy matters. He is the former executive vice president of the Air Transport Association of America, where he represented the interests of the U.S. major airlines industry, particularly related to technical, safety, and security issues. Prior to holding this position, he served as the assistant secretary, Defense Programs, U.S. Department of Energy, where he was responsible for the U.S. nuclear weapons development program, including production, research, testing, safety, and security. He is also a major general, USAF (retired), and held positions of responsibility within NATO, at the Pentagon with the Secretary of the Air Force, and in Vietnam, where he commanded a combat air wing and flew as a fighter pilot. General Hoover was a member or chair of several other NRC study committees and currently serves as chairman of the National Research Council's Aeronautics and Space Engineering Board. He holds a B.S. in engineering from the U.S. Naval Academy, an M.S. in aeronautical engineering from the Air Force Institute of Technology, and is a distinguished graduate of the National War College.

S. MICHAEL HUDSON recently retired as vice chairman of Rolls-Royce North America. After Allison Engine Company was acquired by Rolls-Royce, Mr. Hudson served as president, chief executive officer, chief operating officer, and a member of the board of directors of Allison Engine Company, Inc. Previously, during his tenure at Allison, he served as executive vice president for engineering, chief engineer for advanced technology engines, chief engineer for small production engines, supervisor of the design for Model 250 engines, chief of preliminary design, and chief project engineer in vehicular gas turbines. Mr. Hudson brings insight into propulsion engineering issues, related business issues, and the European perspective on aviation issues. Mr. Hudson served on three other NRC committees.

NANCY G. LEVESON, NAE, received degrees in mathematics, management, and computer science from the University of California, Los Angles (Ph.D. in 1980) and subsequently worked as a computer science professor at the University of California, Irvine. In 1993 she became the Boeing Professor of Computer Science and Engineering at the University of Washington. She is now at the Massachusetts Institute of Technology as a professor of aeronautics and astronautics, where her primary interests lie in software engineering and software and system safety. Dr. Leveson is a member of the board of directors of the Computing Research Association, a member of the Association for Computing Machinery (ACM) Committee on Computers and Public Policy, a consultant to the NASA Aerospace Safety Advisory Panel, a fellow of the ACM, a former member of the advisory committee of the NRC's Division on Engineering and Physical Sciences, and a former member or chair of numerous NRC study committees. She was awarded the ACM 1999 Allen Newell Award for contributions to computer science research and the 1995 AIAA Information Systems award. Dr. Leveson helped the committee address the potential benefits and limitations of using advanced information technology in aircraft and air transportation systems.

RICHARD MARCHI is currently senior vice president, technical and environmental affairs, for the Airports Council International-North America (ACI-NA). He is responsible for overall supervision, direction, and coordination of the staff and activities of the ACI-NA Technical and Environmental Affairs Department. The department provides staff support to five ACI-NA committees: Technical Affairs, Environmental Affairs, Small Airports, Business Information Technologies, and Public Safety and Security. He is also responsible for the development, coordination, and presentation of technical, security, telecommunications, and environmental policies for consideration by the ACI-NA board of directors, as well as preparation of responses to governmental issues of concern to airports, and for developing airport testimony on technical matters. He is the association's focal point representative in preparations for International Civil Aviation Organization (ICAO) technical and environmental matters affecting member airports. Mr. Marchi is an active member of several FAA advisory committees and task forces, including the FAA Free Flight Select Committee, the FAA Research, Engineering and Development Advisory Committee, where he serves as chairman of the Airport Technology Research Subcommittee, and the FAA New Large Aircraft Facilitation Group.

RICHARD R. PAUL is vice president, strategic development, for the Boeing Company's Phantom Works in Seattle. The Phantom Works is Boeing's research and development organization; it is focused on technology development, process improvement, and new product development. Mr. Paul joined Boeing in October 2000 after 33 years with the U.S.

Air Force. During his Air Force career, he served in two Air Force laboratories and in his last assignment he served in a dual-hatted position as the Air Force Technology Executive Officer and commander of the Air Force Research Laboratory. Mr. Paul is a member of the NRC's Air Force Science and Technology Board and a former member of one other NRC study committee. He contributed to this committee expertise and experience from both the Department of Defense and industry.

AMY R. PRITCHETT is an associate professor in the School of Industrial and Systems Engineering and a joint associate professor in the School of Aerospace Engineering at the Georgia Institute of Technology. Her research encompasses cockpit design, including advanced decision aids; procedure design as a mechanism to define and test the operation of complex, multiagent systems such as air traffic control systems; and simulation of complex systems to assess changes in emergent system behavior in response to implementation of new information technology. Dr. Pritchett is the editor of *Simulation: Transactions of the Society for Modeling and Simulation* for the air traffic area; associate editor of the AIAA *Journal of Aerospace Computing, Information, and Communication*; technical program chair for the aerospace technical group of the Human Factors and Ergonomics Society; and co-chair of the 2004 International Conference in Human-Computer Interaction in Aerospace (HCI-Aero). Dr. Pritchett contributed to the committee's investigation of system modeling and automation.

ROBERT J. RAVERA established RJR Aviation, LLC, after retiring from the MITRE Corporation, where he served as vice president for operations in the Center for Advanced Aviation System Development (CAASD). Dr. Ravera now consults on a broad range of aviation and transportation issues, including air traffic control automation; communications, navigation, and surveillance; security; intelligent transportation systems; and infrastructure. At MITRE, he had a key role in overseeing CAASD's work for the Federal Aviation Administration and supported development of MITRE's international aviation program. Other work at MITRE involved Dr. Ravera in modeling and simulation, navigation and surveillance, and other aspects of air traffic control, all of which contributed to the committee's Phase 2 activities.

SANFORD REDERER is president of Aviation Planning & Finance, a small consulting firm. He works as a consultant on airline route and fleet planning, business strategy and marketing programs, aircraft finance, and airport demand management (methods for allocating scarce airfield and facilities access). Clients since 1990 have included airlines, airports, government agencies, airframe and engine manufacturers, and financial institutions in the United States and abroad. Before founding Aviation Planning & Finance, Mr. Rederer was senior vice president-strategic planning at Trans World Airlines, with responsibility for route and fleet planning, merger evaluation and planning, aircraft acquisition, aircraft sales, and alliance development. He served on the staff of the Civil Aeronautics Board in 1977-1980 in several positions, including director of the Bureau of International Aviation, in which position he helped negotiate significant liberalization of the Bermuda 2 bilateral air services agreement. Mr. Rederer earned an A.B. degree in economics from Hamilton College and an M.A. in economics from the University of California, Berkeley, and is a veteran of the U.S. Army. Mr. Rederer brings to the committee important experience in airline economics, planning, and management.

HERBERT H. RICHARDSON, NAE, is director of the Texas Transportation Institute and associate vice chancellor for engineering at the Texas A&M University system. He is also Regents Professor and Distinguished Professor of Engineering at the university. From 1991 to 1993 he was chancellor of the Texas A&M University System. Before joining Texas A&M in 1984, he was associate dean of engineering at the Massachusetts Institute of Technology (MIT), where he began his academic career in 1955. He was head of MIT's Mechanical Engineering Department from 1974 to 1982. On leave from MIT, he was chief scientist for the U.S. Department of Transportation from 1970 to 1972. He has served on many NAE and NRC committees, including the Council of the NAE and the NRC Governing Board. He chaired TRB's executive committee, the Committee for the Critique of the Federal Research Program on Magnetic Levitation Systems, and the Committee for a Study of the Railroad Tank Car Design Process. He was co-chair of the Committee for the Study of Geometric Design Standards for Highway Improvements and vice chair of the Committee for a Review of the National Automated Highway System Consortium Research Program. Dr. Richardson earned a Ph.D. in mechanical engineering from MIT and brings a broad transportation policy and technology perspective to the committee.

RUSSELL D. SHAVER III is a senior policy analyst with RAND. He has worked on a wide array of topics, including transportation security and the future of the FAA and the national airspace system. Dr. Shaver helped the committee understand the strengths and weaknesses of existing models as they relate to predicting the future performance of the national airspace system.

DAVID D. WOODS is a professor in the Institute for Ergonomics at the Ohio State University. He is an expert in cognitive engineering, investigating problems such as human error, how complex systems fail, how to make intelligent systems team players, and automation surprises in application areas such as space operations and automated flight decks. He has received awards for research on integrated pattern displays (Ely award for best paper in Human Factors, 1994), on cockpit automation (Laurel Award from *Aviation*

Week and Space Technology in 1995), and on cognitive engineering (the Kraft Innovators award from the Human Factors and Ergonomic Society in 2002). He is a former president and fellow of the Human Factors and Ergonomic Society and a fellow of the American Psychological Society and the American Psychological Association. He served on two other NRC committees. Dr. Woods's expertise helped the committee address automation issues.

D

Propulsion Taxonomy: Comments on Propulsion Fundamentals

The purpose of this taxonomy, which is a generalization of Zwicky's idea (Zwicky, 1959), is to suggest that there are many unexplored ideas in the realm of aircraft propulsion. As with any taxonomy, not all the ideas are fruitful. Some can be immediately rejected as violating physical principles, and other ideas may not be capable of developing a suitable energy density. Nevertheless, a simplified taxonomy of aircraft propulsion provides a means of sorting and defining a large number of propulsion devices, some of which are in use, some of which may be practical following an adequate level of research, and some of which may never be practical.

Assuming an isentropic compression and expansion phase of the cycle, the committee identified 10 fundamental cycles that use fuel to provide thrust, shaft power, or electricity as an output (see Table D-1). Nine of these are thermodynamic in nature and the tenth converts fuel directly to electricity. This is the fuel cell. Batteries have intentionally been omitted because considerable development seems to be necessary before this power source can be applied to the propulsion of common as opposed to niche airplanes.

TABLE D-1 Fundamental Thermodynamic Cycles (nonregenerative)[a]

	Heat Absorption	Heat Rejection	Comment
1	Isothermal	Isothermal	Carnot cycle
2	Isothermal	Isovolume	
3	Isothermal	Isobaric	
4	Isovolume	Isothermal	
5	Isovolume	Isovolume	Otto cycle
6	Isovolume	Isobaric	
7	Isobaric	Isothermal	
8	Isobaric	Isovolume	Diesel cycle
9	Isobaric	Isobaric	Brayton cycle

NOTE: The tenth fundamental source of power is a fuel cell, which converts a fuel directly into electricity.

[a]Compression and expansion processes are assumed to be isentropic.

Alternating-current electric motors are relatively heavy. New magnetic materials provide lighter weight direct-current motors. This may or may not be the most immediate application of power from fuel cells. The heat added by a resistance heater is confined to a thermal boundary layer, which like all boundary layers is quite thin. Any attempt to emulate volumetric heating requires a relatively dense distribution of resistive elements and thus a sizable pressure drop. For this reason a fuel cell seems unlikely to be used to power a resistance heater to replace the combustor in an aircraft engine. However, the power from a fuel cell might be used for a volumetric heating process using a plasma, a laser, or a microwave breakdown process. If room-temperature superconductivity becomes a reality, a rotating electromagnetic wave in the nacelle could be used to suspend and drive a fan. The power for this arrangement could conceivably be derived from a fuel cell of the future.

The first column of a propulsion taxonomy (see Table D-2) could be defined by 10 items, the 9 cycles given in Table D-1 plus fuel cells. The second column would have two items: continuous and intermittent. Thus,

Cycle	*Operation*
1 Isothermal-isothermal	1 Continuous
2 Isothermal-isovolume	2 Intermittent
3 Isothermal-isobaric	
4 Isovolume-isothermal	
5 Isovolume-isovolume	
6 Isovolume-isobaric	
7 Isobaric-isothermal	
8 Isobaric-isovolume	
9 Isobaric-isobaric	
10 Fuel cell	

TABLE D-2 Matrix Summary of Propulsion Taxonomy

Cycle[a]	Operation	Pressurization	Energy release	Propulsion
1 Isothermal-isothermal	1 Continuous	1 Mechanical	1 Oxidation (combustion)	1 Propeller
2 Isothermal-isovolume	2 Intermittent	2 Self-pressurized	2 Electrochemical (fuel cell)	2 Turbofan
3 Isothermal-isobaric		3 Both	3 Photochemical (photosynthesis)	3 Turbofan and afterburner
4 Isovolume-isothermal			4 Photoelectric (solar cell)	4 Turbojet
5 Isovolume-isovolume			5 Photodirect (laser/electro-magnetic heating)	5 Ramjet
6 Isovolume-isobaric				6 Rocket
7 Isobaric-isothermal				7 Turbojet and afterburner
8 Isobaric-isovolume				8 Pulse-jet (pulse detonation engine)
9 Isobaric-isobaric				
10 Fuel cell				

[a]Cycle name describes methods of heat absorption and heat rejection, respectively.

Selecting one item from the first column and one from the second column defines, in an elemental way, a family of 20 power plants. However, the continuous isothermal and isovolume cycles can be immediately ruled out as physically impossible because in the steady state the fluid is forced to move uphill against the total pressure gradient. This leaves 12 potential power plants (10 intermittent and 2 continuous: isobaric-isobaric and fuel cell). The next column of the taxonomy is pressurization:

Pressurization

1 Mechanical
2 Self-pressurized
3 Both

These three entries create a total of 36 options (30 intermittent and 6 continuous). Consider next energy release processes from fuel. The terms in parentheses are common names for the processes.

Energy Release

1 Oxidation (combustion)
2 Electrochemical (fuel cell)
3 Photochemical (photosynthesis)
4 Photoelectric (solar cell)
5 Photodirect (laser/electromagnetic heating)
Etc.

Recombination of excited molecular and atomic states, free radicals, and antimatter recombination are included in the term "etc." but are not further considered because of the problems of storing the fuels. Nuclear energy release is excluded because of the weight of shielding material and radioactive hazards. With this exclusion, we now have five more options, which creates a total of 180 possible devices. Photochemical, photoelectric, and photodirect systems arguably possess a low energy density and in all likelihood will only find niche applications. Setting aside low-power-density processes reduces the number of options to 72 (60 intermittent and 12 continuous).

Propulsion mechanism is the next area for consideration.

Propulsion

1 Propeller
2 Turbofan
3 Turbofan + afterburner
4 Turbojet
5 Ramjet
6 Rocket
7 Turbojet + afterburner
8 Pulse-jet (pulse detonation engine)

These eight items provide one intermittent propulsion option and seven continuous propulsion options, one of which (the propeller) can also be used with intermittent cycles. This results in 204 options ($2 \times 60 + 7 \times 12$), which is quite a large number.

COMMENTS

Both the constant temperature and constant volume heat absorption (or combustion) cycles are usually intermittent (pulsed). Intermittent processes have the potential to operate at higher temperatures and reasonable wall temperatures because the wall can be cooled during the part of the cycle when no heat is applied. However, in practice some of this heat is lost and reduces efficiency. Further, while intermittent cycles have a high efficiency per pulse, the average efficiency is lower because power is available only part of the time.

The fuel cell could require a precompression of fuel depending on the fuel storage mechanism. The fuel cell has the potential to be isothermal and is thus quite attractive. Coupled to a superconducting electric motor, it could have a very high efficiency with low hydrocarbon emissions. The water vapor emissions can be reduced to zero, though at some expense in weight.

System weight and thermal efficiency are the fundamental system issues. The three curves in Figure D-1 show cycle efficiency as a function of the ratio of the heat rejection temperature (T_2) to the heat absorption temperature (T_1). The Carnot cycle is clearly the most efficient, with the Otto cycle second and the Brayton cycle the least efficient. The comparison is somewhat misleading, however, since efficiencies for the Carnot and Otto cycles are per pulse and not the average over the cycle. Carnot and Otto cycle efficiencies are somewhat lower in practice. An additional curve in Figure D-1 shows the efficiency of the Otto cycle if power is produced during only half the cycle.

The isobaric cycles can be continuous or intermittent since the pressure falls during combustion processes. Each process has its limitations. For example, the efficiency of the Brayton cycle is limited by material limits. That is, the compressor exit temperature is limited by high temperature material properties. Nevertheless, the continuous or open Brayton cycle provides a high power density and a relatively simple structure.

Relaxing the constraint on adiabatic compression and expansion (e.g., by using regeneration) almost doubles the number of options. For example, it may be possible for heat removed during compression to be added during expansion in the turbine. With the exceptions described above, the current state of the art for heat exchanger technology is based on surface heat removal, not volume heat removal. The thermal boundary layer is thin even in an axial flow compressor, meaning the heat exchanger is likely to be heavy and create additional pressure losses. Hence until a volumetric heat exchange process is invented, the regenerative part of a modified Brayton cycle seems impractical.

Also, a cooling system upstream of the compressor could inject mist into the air stream. This would have two advantages: The stream would be cooled by evaporation and the total pressure would be increased. This is Ascher Shapiro's aerothermo compressor. It might even be possible to inject the mist ahead of each compressor rotor and stator blade. In principle the cooling could be adjusted such that the temperature remains constant during compression. In practice, it may be possible to increase the effective compressor efficiency while reducing compressor outlet temperature by 50 to 150 °F. Another option would be to allow combustion in a suitably designed turbine stage, so that heat could be added in a more nearly isothermal fashion. This modified Brayton cycle seems attractive enough to warrant further research.

In summary, there seem to be a large number of alternative propulsion schemes. However, at present few alternatives seem to be practical; only a very few, including modified Brayton cycle engines, seem to warrant more than passing attention.

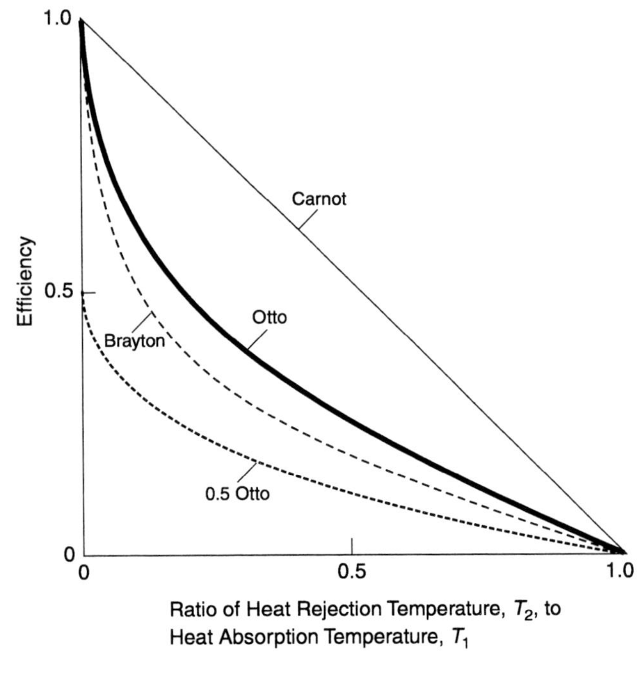

FIGURE D-1 Thermal efficiency of the Otto, Brayton, and Carnot cycles.

REFERENCE

Zwicky, F. 1959. Future Prospects of Jet Propulsion. Volume XII, Section L, Jet Propulsion Engines: High Speed Aerodynamics and Propulsion. Princeton, N.J.: Princeton University Press.

E

Four Levels of Models

This appendix provides an illustrative example of four levels of models, running from specific to general, that could be used to support modeling and analysis of the overall system. The outputs of each lower-level model support the higher-level models. An airport landing aide has been selected as the subject of the lowest level model in this example.

Level 1. Assessing the ability of a Local Area Augmentation System (LAAS) to support landings at a particular airport. Characteristic inputs to a Level 1 model of technologies would be LAAS signal structure, the impact of terrain on signal propagation, and the capabilities of aircraft avionics systems (including signal processing and inertial navigation systems). The model's output would be aircraft position accuracy vis-à-vis the runway and what this would mean for landing the airplane in various weather conditions. Level 1 models can also support detailed analyses of human performance through human-in-the-loop simulations or through detailed agent-based simulations using computational human performance models.

Level 2. Assessing airport landing and takeoff capacity in various weather conditions. LAAS performance would be just one input to this model. Other inputs would include the sequencing of aircraft, how individual aircraft are outfitted, and other automation aides—for example, the Center TRACON Automation System (CTAS) and wake vortex constraints (see Figure E-1). LAAS performance is intended to increase overall airport landing capacity, but the ability to do so is sensitive to runway configuration, the actual weather conditions, aircraft equipage, controller workload, etc. The output of this Level 2 model would characteristically be a set of throughput numbers that are associated with different weather states, although this would obscure the stochastic nature of the model results (see Figure E-2). A Level 2 model might also be used to predict airport congestion. The inputs to such a model would include scheduled arrivals and departures as well as available taxiways and gates. Models of this sort are often used to design and evaluate airport improvements, including new runways and expanded passenger facilities. Figure E-3 shows an example of a model output being used to balance the availability of arrival gates against runway capacity.[1]

Level 3. Assessing delays across the air transportation system. Inputs include airline schedules (real and projected, based on estimated future demand), airport capacities, aircraft routes between city-pairs, and en route air traffic control sector capacities (see Figure 3-1). The outputs of these models are typically average delays across the entire system, the distribution of such delays by airport, and the mitigation of delays as a function of various changes to the inputs (including airport capacity). The actual model outputs are stochastic, but the stochastic nature of these outputs is often ignored. Other outputs, often ignored, include the delays encountered at specific airports over the course of the day (see Figure E-4). These "ignored" outputs are often important because they help explain the character of the results and the causes for the delays. However, delays at even a single airport may be difficult to calculate; a systemwide model is required to accurately predict traffic levels at an individual airport over the course of a day.

Level 4. Assessing the impact of inadequate air transportation system capacity on the national economy. In contrast to the above models, which were primarily simulations of performance at different levels of detail, this model is prima-

[1]Airports are complicated systems, and efficient gate utilization is one of the largest challenges airlines face during busy hours. Airlines and airports use a variety of models (e.g., large-scale simulations like TAAM or special-purpose gate assignment models with sophisticated artificial intelligence programming) to judge capacity constraints. Similarly, the FAA and its contractors use complicated models to judge runway throughput under a variety of conditions.

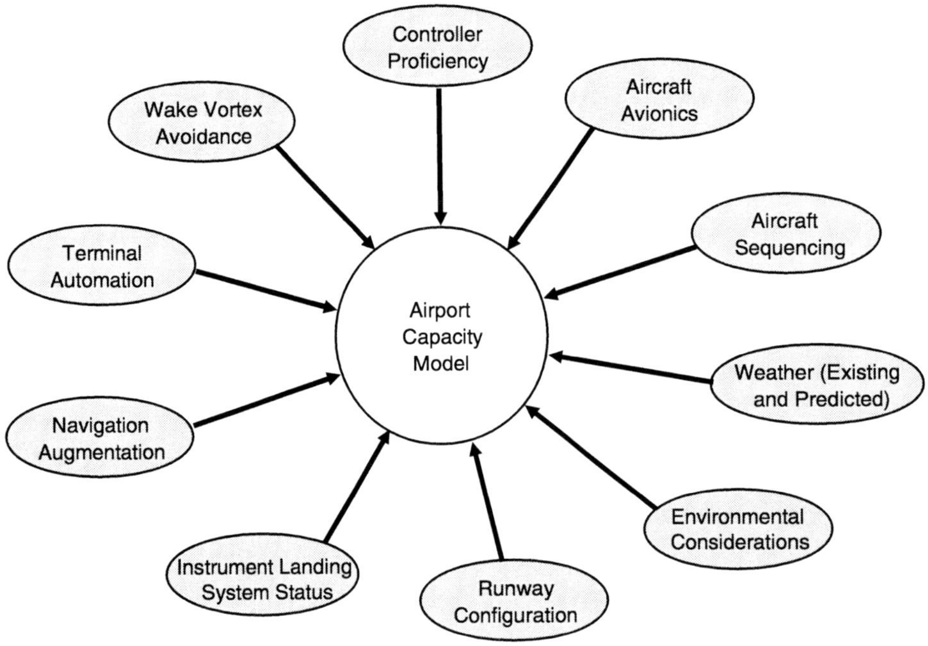

FIGURE E-1 Generic inputs for a model of airport capacity.

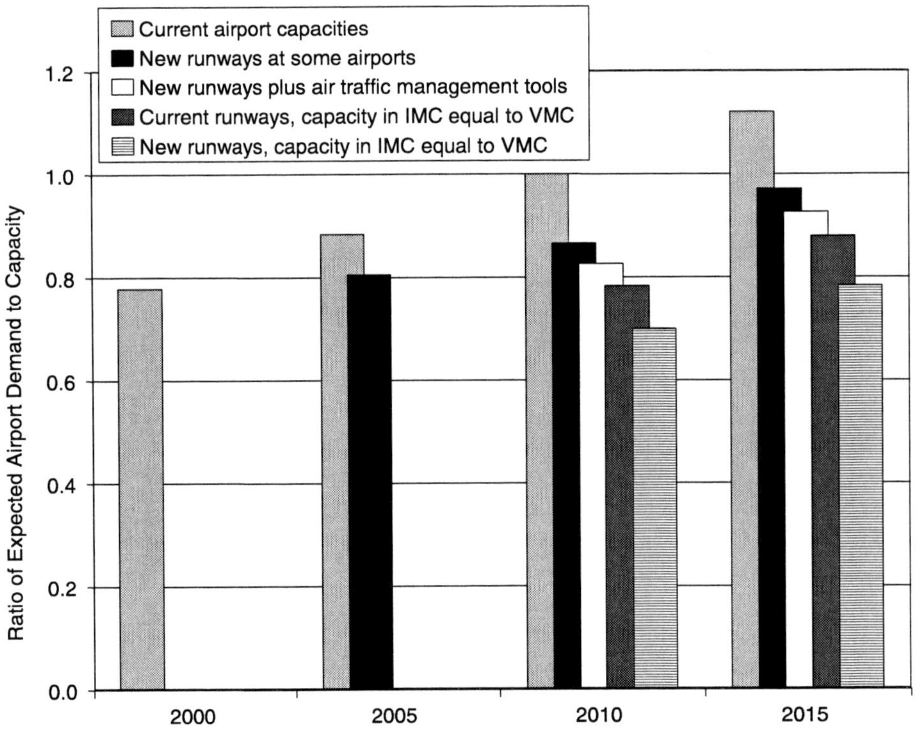

FIGURE E-2 Ratio of expected demand (in terms of landings and takeoffs from an airport) to airport throughput capacity as a function of time (2000 to 2015) and planned airport and terminal area improvements for the 31 largest U.S. airports. IMC, instrument meteorological conditions; VMC, visual meteorological conditions.

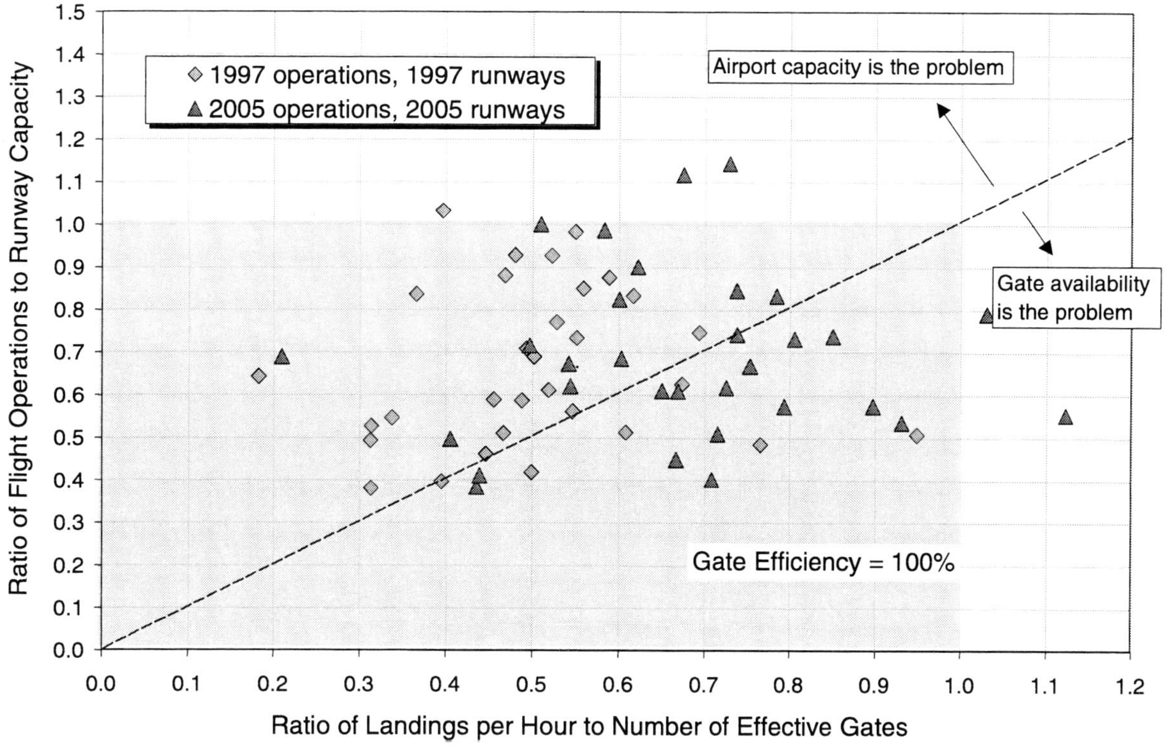

FIGURE E-3 Influence of runway capacity and number of available gates on throughput at the 30 busiest airports in the United States in visual meteorological conditions. Above the dashed line, landing rates are the primary constraint on airport throughput; below the line, gates are the primary constraint.

rily analytic. It has as its inputs the elasticity of passenger (and freight) demand to changes in both ticket price and convenience (essentially another cost to the passenger). Convenience includes frequency of departures to the desired city, the amount of time spent in the airport (which may be driven by aircraft delays as well as security measures at the airport), and the expectation that passengers will actually arrive at their destination without significant delay. This latter concern usually equates to whether a flight will be cancelled or passengers will miss their connections at a hub airport. Also included in such models is an expectation of how the airlines will manage increased demand and the potential for alternative transportation modes that might supplant or complement air travel. Figure E-5 provides an example of the output of such models.

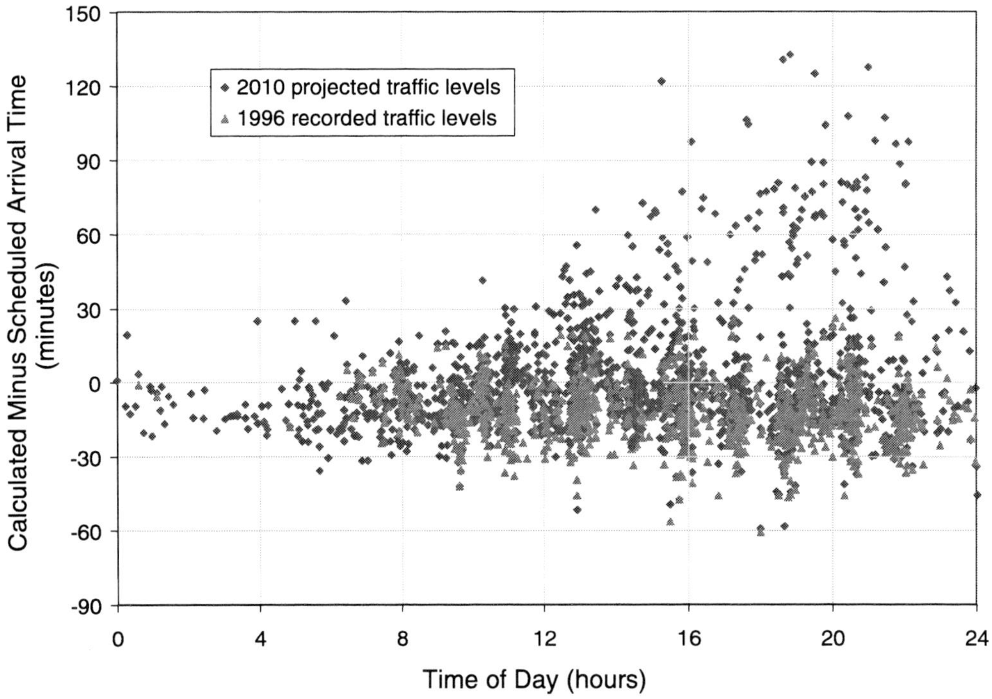

FIGURE E-4 Impact of traffic growth on scheduling predictability at a major U.S. airport in visual meteorological conditions, based on a comparison of scheduled arrival times versus computed arrival times for 1997 (real data) and 2010 (projected data, assuming traffic increases 2.3 percent per year). The large delays occurring in 2010 are not caused by this airport's planned schedule. The planned arrival rate never exceeds the airport's acceptance rate, and more planes are not scheduled to arrive at the end of the day than at other times during the day. Delays primarily reflect the accumulated effect of delays at other airports, which result in late arrivals.

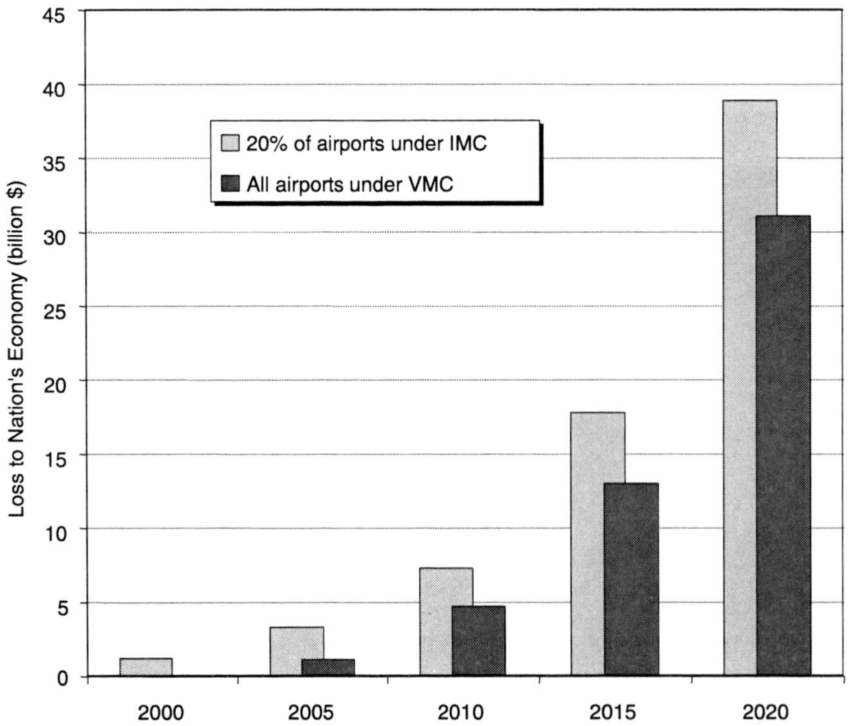

FIGURE E-5 Economic losses caused by undercapacity at U.S. airports, assuming that improvements to the air transportation system occur as scheduled.

F

Acronyms and Abbreviations

BWB	blended-wing-body
CO_2	carbon dioxide
CTAS	Center TRACON Automation System
dB	decibel
DESIDE	Discrete-Event Simulation Interactive Development Environment
DNL	day-night average sound level
DPAT	Detailed Policy Assessment Tool
EDMS	Emissions and Dispersion Modeling System
FAA	Federal Aviation Administration
IHPTET	Integrated High Performance Turbine Engine Technology (program)
IMC	instrument meteorological conditions
LAAS	Local Area Augmentation System
L/D	lift-to-drag ratio
LMI Net	Logistics Management Institute network simulation model
MAGENTA	Model for Assessing Global Exposure to Noise from Transport Aircraft
MEMS	microelectromechanical systems
MTOW	maximum takeoff weight
NAE	National Academy of Engineering
NASA	National Aeronautics and Space Administration
NASTRAN	NASA Structural Analysis program
NO_x	oxides of nitrogen
NRC	National Research Council
RTCA	The name of a not-for-profit corporation. (The name is no longer an acronym.)
SAGE	System for assessing Aviation's Global Emissions
SATS	Small Aircraft Transportation System

TAAM	Total Airspace and Airport Modeler
TRACON	Terminal Radar Control Facility
UAV	uninhabited air vehicle
VMC	visual meteorological conditions
VTOL	vertical takeoff or landing